AMS SHORT COURSE LECTURE NOTES
Introductory Survey Lectures
published as a subseries of
Proceedings of Symposia in Applied Mathematics

Proceedings of Symposia in
APPLIED MATHEMATICS

Volume 61

Trends in Optimization

American Mathematical Society
Short Course
January 5–6, 2004
Phoenix, Arizona

Serkan Hoşten
Jon Lee
Rekha R. Thomas
Editors

American Mathematical Society
Providence, Rhode Island

LECTURE NOTES PREPARED FOR THE
AMERICAN MATHEMATICAL SOCIETY SHORT COURSE
TRENDS IN OPTIMIZATION
HELD IN PHOENIX, ARIZONA
JANUARY 5–6, 2004

The AMS Short Course Series is sponsored by the Society's Program Committee for National Meetings. The series is under the direction of the Short Course Subcommittee of the Program Committee for National Meetings.

2000 *Mathematics Subject Classification.* Primary 05A15, 13P10, 13F55, 42B10, 52A41, 52B20, 52C07, 90B80, 90Cxx.

Library of Congress Cataloging-in-Publication Data

American Mathematical Society. Short course (2004 : Phoenix, Ariz.)
Trends in optimization : American Mathematical Society Short Course, January 5–6, 2004, Phoenix, Arizona / Serkan Hoşten, Jon Lee, Rekha R. Thomas, editors.
p. cm. — (Proceedings of symposia in applied mathematics, ISSN 0160-7634 ; v. 61, AMS short course lecture notes)
Includes bibliographical references and index.
ISBN 0-8218-3584-X (alk. paper)
1. Mathematical optimization—Congresses. I. Hoşten, Serkan, 1969– II. Lee, Jon, 1960– III. Thomas, Rekha R., 1967– IV. American Mathematical Society. V. Title. VI. Proceedings of symposia in applied mathematics ; v. 61. VII. Proceedings of symposia in applied mathematics. AMS short course lecture notes.

QA402.5.A4534 2004
519.6—dc22 2004059477

10 9 8 7 6 5 4 3 2 1 09 08 07 06 05 04

Contents

Preface

The American Mathematical Society sponsors a series of *Short Courses*, each of which consists of a set of survey lectures on a single theme of pure and applied mathematics. Initiated in 1973, these two-day courses occur immediately preceding the Joint Mathematics Meetings held in January as well as at some summer meetings.

On January 5-6, 2004, the AMS Short Course *Trends in Optimization 2004* took place in Phoenix, Arizona. Optimization has not been the focus of a recent AMS Short Course. Past Short Courses on related themes consist of: (i) *Operations Research*, Washington D.C., 1975, (ii) *Applied Combinatorics*, Kalamazoo, 1975, and (iii) *Operations Research*, Duluth, 1979. There have been impressive advances in optimization since then, and our goal was to showcase some of the exciting more recent work.

Our Short Course consisted of seven 75-minute lectures given by leaders in the field of optimization. It is impossible to give a thorough cross-section of research in optimization in a two-day course. Rather, we chose seven exciting areas to focus on, with a clear bias toward the *discrete* side of optimization, somewhat reflecting the interests of the organizers. This volume comprises edited notes prepared by our lecturers.

Optimization is concerned with the efficient computation of the supremum of an *objective function* f whose domain is restricted to some set of *feasible solutions* S. Assumptions about the function f (linear, convex, continuous, differentiable, etc.) and the set S (a hypercube, a convex set, a polyhedron, the integer lattice points in a polyhedron, the set of symmetric positive semidefinite matrices, etc.) lead to structural results and/or efficient algorithms. When the set S is a subset of the power set of a finite set, we are in the domain of combinatorial optimization.

No sophisticated background is needed in order to read the lecture notes in this volume. Many of them are self-contained and all provide extensive references. However, basic knowledge of linear programming duality (which appears in almost all chapters), the geometric combinatorics of polyhedral sets (extreme points and rays, faces, valid inequalities, etc.), and the geometry of integer programming would be useful. There are many nice books devoted to each topic above, but most cover more than the introductory material. A good starting point where the reader can find the needed material in a condensed but very readable form is Chapter 0 of Jon Lee's textbook *A First Course in Combinatorial Optimization* (Cambridge University Press, 2004).

Karen Aardal (Georgia Institute of Technology) in her lecture *Lattice basis reduction in optimization,* describes Lovasz's fundamental algorithm for producing a short vector in a lattice by lattice basis reduction, and H.W. Lenstra's use (in the early 1980's) of this idea in his polynomial-time algorithm for integer programming in fixed dimension. This lecture note is one of the most lucid presentations of the material, and it also contains pointers to the developments on the practical side with an eye towards computational tools.

In his lecture *Polyhedral methods in discrete optimization,* Alper Atamtürk (University of California at Berkeley) surveys some of the powerful cutting-plane techniques which, in conjunction with modern computational tools for linear programming, have enabled the solution of large mixed integer linear programs. Particular attention is paid to the so-called *lifting* techniques for producing valid inequalities from simpler polyhedra such as knapsack polyhedra. These ideas are showcased in an application in robust combinatorial optimization.

In Gérard Cornuéjols's (Carnegie Mellon University) lecture *Graphs and combinatorial optimization,* we see the rich interplay between linear programming duality and combinatorial optimization problems arising from graphs and related structures. Important ideas such as packing and covering properties, perfectness and total dual integrality, are presented in a compact and coherent way. Most importantly, this lecture note contains a very readable outline of the recent proof of the Strong Perfect Graph Theorem by Chudnovsky, Robertson, Seymour and Thomas.

Linear programming duality has been a powerful tool for proving combinatorial min-max theorems. However, there is no commonly agreed upon way of extending these results to an integer programming duality. Jean-Bernard Lasserre (LAAS-CNRS, Toulouse), in his lecture *Integer-programming duality,* describes his recent and very promising work on a natural duality for integer programming based on generating functions and counting formulas. Ideas from Fourier analysis (Laplace transform) and complex analysis (residues) play a prominent role in this new and exciting direction.

Many combinatorial optimization problems are NP-hard, and are thus unlikely to admit exact solutions in polynomial time. David Shmoys (Cornell University), in his lecture *The design and analysis of approximation algorithms: Facility location as a case study,* surveys methods for the efficient computation of near-optimal solutions. Specifically, he focuses on algorithmic paradigms that yield efficient algorithms for producing a solution whose objective value is guaranteed to be within a given factor of the optimal objective value. The facility location problem is used as a relatively simple example to illustrate the fundamental ideas and directions.

In his lecture *Algebraic recipes for integer programming,* Bernd Sturmfels (University of California at Berkeley), describes how methods of commutative algebra and algebraic combinatorics can be used to successfully attack integer programming problems. In particular, Gröbner bases play a central role in algorithmic theory and practice. Moreover, techniques based on short rational functions are bringing new insights, such as in computing the integer programming gap.

Steve Wright (University of Wisconsin) surveys the terrain of *Nonlinear and semidefinite programming.* The techniques used in this subfield of optimization tend to be very different than those typically used in discrete optimization. The area, though only represented by one lecture, is a broad and exciting field. The development of efficient interior-point algorithms for linear and semidefinite-programming problems, and connections with approximation algorithms for combinatorial optimization problems, have generated considerable interest within the discrete optimization community.

We are grateful to Annalisa Crannell who helped us with our proposal to the AMS, and Wayne Drady and the rest of the AMS staff who helped us every step of the way. We are also very thankful to everyone who helped us review the chapters: Jesús De Loera, Christian Haase, Tyrell McAllister, Alex Milowski, Edwin O'Shea, Will Traves, Andreas Wächter and Rudy Yoshida did excellent jobs under a tight schedule.

Serkan Hoşten, *San Francisco State University, San Francisco*
Jon Lee, *IBM T.J. Watson Center, Yorktown Heights*
Rekha Thomas, *University of Washington, Seattle*

March 2004

Proceedings of Symposia in Applied Mathematics
Volume **61**, 2004

Lattice basis reduction in optimization: Selected topics

Karen Aardal

In memory of my father.

ABSTRACT. Lattice basis reduction has proven to be a powerful tool in many areas of discrete mathematics, such as cryptography and integer programming. We will review some of these results with an emphasis on Lovász' basis reduction algorithm, H.W. Lenstra's algorithm for integer programming, and some related algorithms. These integer programming algorithms have in common that they enumerate lattice hyperplanes, i.e., hyperplanes in which the whole lattice is contained. By using basis reduction, Lenstra proved that for fixed dimension the number of hyperplanes we need to consider is bounded by a function that depends on the dimension only. This result not only resolved an important open question, but introduced new tools to integer optimization.

1. Introduction and motivation

In our lecture we mainly focus on the following *integer linear feasibility problem.* Let $P = \{\boldsymbol{x} \in \mathbb{R}^n \mid \boldsymbol{Ax} \leq \boldsymbol{d}\}$, where the $m \times n$ matrix $\boldsymbol{A}$ and the m-vector $\boldsymbol{d}$ are given by integer input. Assume P is bounded and full-dimensional.

(1.1) Does there exist a vector $\boldsymbol{x} \in P \cap \mathbb{Z}^n$?

This problem is related to the *integer linear optimization problem*, where we want to find a solution that maximizes or minimizes a given objective function $\boldsymbol{cx}$, where $\boldsymbol{c}$ is an n-dimensional row vector. Branch-and-bound has proven to be a very successful method to solve integer linear optimization problems, especially when combined with cutting planes, and it is implemented in several commercial optimization packages. The key component of branch-and-bound is to solve the linear relaxation, i.e., we maximize (or minimize) $\boldsymbol{cx}$ over P. If the optimal solution $\bar{\boldsymbol{x}}^*$ to the linear relaxation is integer, then we are done; otherwise we choose a non-integral component of $\bar{\boldsymbol{x}}^*$, say x_j with value f_j and create two subproblems by adding the constraints $x_j \leq \lfloor f_j \rfloor$, and $x_j \geq \lceil f_j \rceil$ respectively to P. This is the part of the algorithm that is called "branching". Each of the subproblems is again treated in the same fashion.

2000 *Mathematics Subject Classification.* Primary 90C10; Secondary 11H06, 68Q25.
Key words and phrases. Integer optimization, Lattice basis reduction, Branching on hyperplanes.
The author was supported in part by NSF grants DMI-0100020 and DMII-0121495.

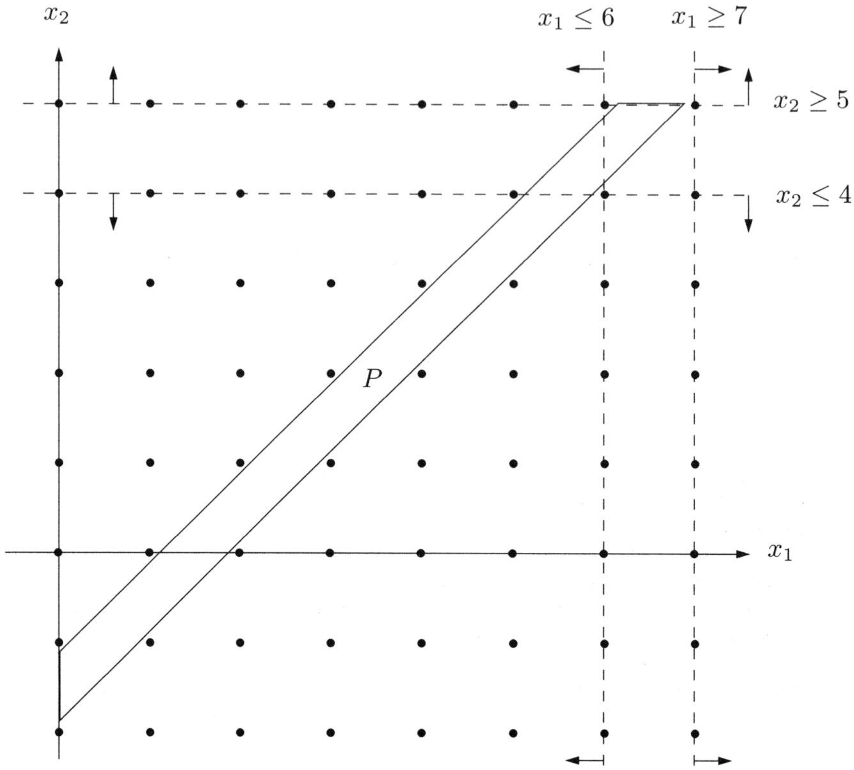

FIGURE 1. The polytope P of Example 1.1, and some constraints added in branch-and-bound

We do not continue branching at a subproblem if the linear relaxation corresponding to that subproblem is infeasible, or if the optimal solution to the relaxation is integer, or if the value $\boldsymbol{c}\bar{\boldsymbol{x}}^*$ is "worse" than the value of a known integer solution (this is the "bounding" part of the algorithm).

Sometimes, though, branch-and-bound does not work well. In some of these cases it seems as if the linear relaxation does not provide useful information, and in other cases we simply do not understand why it does not work. From the theoretical point of view we know that branch-and-bound can perform arbitrarily bad even in dimension $n = 2$. This is illustrated in Example 1.1. A natural question to ask is whether there exists an algorithm for solving Problem 1.1 that has a polynomial running time. Since Problem 1.1 is NP-complete [**7, 21**] we do not expect to find a polynomial algorithm to solve it, but we could still hope for an algorithm that is polynomial for fixed dimension n. Such an algorithm was developed by H. W. Lenstra, Jr. [**25**]. His algorithm is based on lattice basis reduction and linear transformations. The basic principle behind this algorithm is also sketched in Example 1.1 below.

EXAMPLE 1.1. Consider the polytope

$$P = \{x \in \mathbb{R}^2 \mid 5x_1 - 5x_2 \leq 6,\ 5x_1 - 5x_2 \leq 9,\ x_1 \geq 0,\ x_2 \leq 5\}.$$

Does P contain an integer vector?

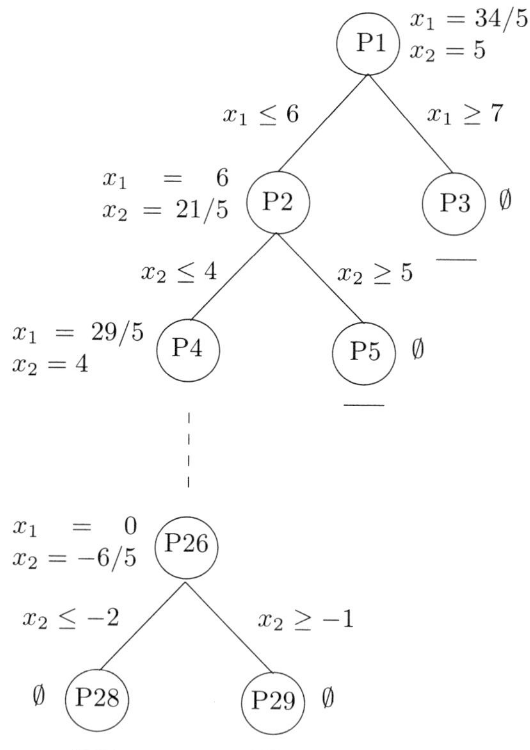

FIGURE 2. The branch-and-bound search tree.

If we solve this feasibility problem by branch-and-bound, we first need to introduce an objective function. Let us choose

$$\max z = x_1 + x_2 \,.$$

If we maximize z over P, i.e., if we solve the linear relaxation of our problem, we obtain the vector $(x_1, x_2)^T = (6\frac{4}{5}, 5)$. We illustrate P and some of the constraints (dashed lines) added during branch-and-bound in Figure 1, and the search tree corresponding to the branch-and-bound procedure in Figure 2. Since $(6\frac{4}{5}, 5)$ is not an integer vector, we create two branches at the root node of our search tree: one corresponding to $x_1 \leq 6$ (subproblem P2), and one corresponding to $x_1 \geq 7$ (subproblem P3). Again, solving the linear relaxation corresponding to subproblem P2 gives the solution $(x_1, x_2)^T = (6, 4\frac{1}{5})$, whereas subproblem P3 is infeasible. Branch-and-bound continues in a similar fashion until subproblems P28 and P29, where all nodes of the search tree are pruned and it is finally verified that P does not contain any integer vector.

An alternative way to solve the problem is as follows. First we observe that all integer points in $\mathbb{R}^n$ are contained in countably many parallel real hyperplanes (see the dashed lines in Figure 3). We can direct these hyperplanes in infinitely many ways. Consider the integer direction $\boldsymbol{d} = (-1, 1)$, and compute the integer width of P in the direction $\boldsymbol{d}$:

$$W_I(P, \boldsymbol{d}) = \lfloor \max\{\boldsymbol{dx} \mid \boldsymbol{x} \in P\} \rfloor - \lceil \min\{\boldsymbol{dx} \mid \boldsymbol{x} \in P\} \rceil + 1 = -2 - (-1) + 1 = 0\,.$$

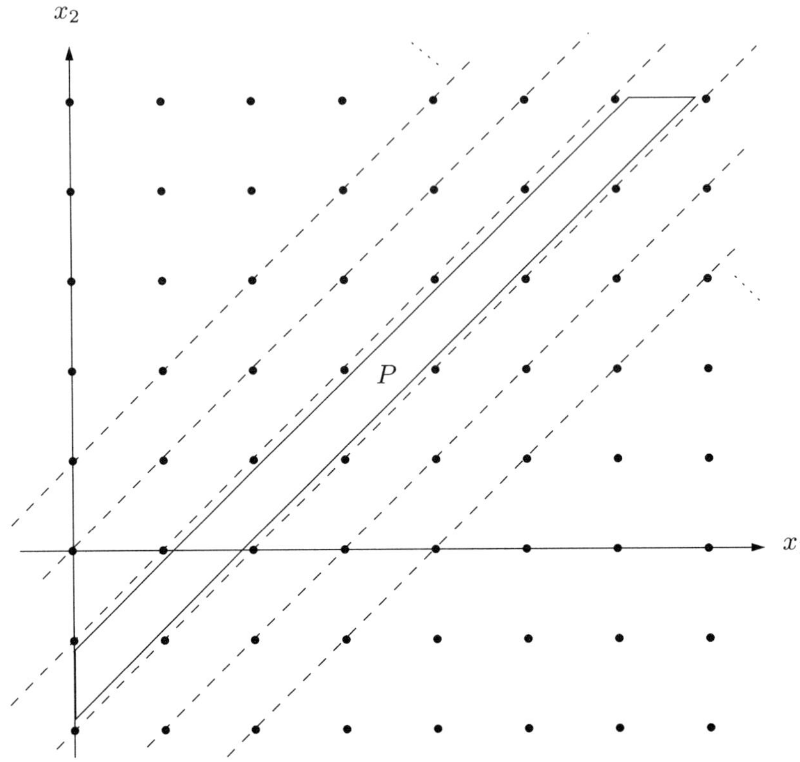

FIGURE 3. The lattice is contained in countably many real hyperplanes.

The number of hyperplanes intersecting P is equal to $W_I(P, \boldsymbol{d})$, which in our example is equal to 0. Since all integer points in $\mathbb{R}^2$ are contained in these "lattice hyperplanes", and none of the hyperplanes is intersecting P, we immediately have a certificate that P does not contain an integer vector. □

Example 1.1 is extreme in order to illustrate that branch-and-bound can fail miserably already in dimension $n = 2$, i.e., even in dimension $n = 2$ there is no bound depending only on n on the size of the search tree. If we had made the polytope P arbitrarily "long", the search tree would have been arbitrarily deep. The second approach illustrates the technique of "branching on lattice hyperplanes". Lenstra's [**25**] algorithm is based on this principle. Before we get into the more technical part of this presentation we want to add some more history and motivation behind this algorithm.

A convex set K is symmetric about the origin if $\boldsymbol{x} \in K$ implies that $-\boldsymbol{x} \in K$.

THEOREM 1.2 (Minkowski [**28**]). *If K is any convex set in $\mathbb{R}^n$ that is symmetric about the origin and has a volume greater than 2^n, then K contains a nonzero vector $\boldsymbol{x} \in \mathbb{Z}^n$.*

Most of the polytopes we are studying in integer programming are not symmetric about the origin, so an interesting question is whether we can make some statement about conditions on the "shape" of a polytope P if P does not contain an integer point. We will state such a result for convex sets.

DEFINITION 1.3. Let K be a full-dimensional closed convex set in $\mathbb{R}^n$ given by integer input. The *width of K along the nonzero vector $\boldsymbol{d}$* is defined as $W(K, \boldsymbol{d}) = \max\{\boldsymbol{dx} \mid \boldsymbol{x} \in K\} - \min\{\boldsymbol{dx} \mid \boldsymbol{x} \in K\}$. The *integer width* of K along the nonzero integer vector $\boldsymbol{d}$, $W_I(K, \boldsymbol{d})$, is defined as

$$W_I(K, \boldsymbol{d}) = \lfloor \max\{\boldsymbol{dx} \mid \boldsymbol{x} \in K\} \rfloor - \lceil \min\{\boldsymbol{dx} \mid \boldsymbol{x} \in K\} \rceil + 1 .$$

THEOREM 1.4 (Khinchine's flatness theorem [**22**]). *Let K be a closed convex set given by integer input. Then, either K contains an integer point, or there exists a nonzero, integer vector $\boldsymbol{d}$ such that $W(K, \boldsymbol{d}) \leq f(n)$.*

Currently the best asymptotic bounds on $f(n)$ are given in [**5**]. Tight bounds seem to be unknown already in dimension 3.

To appreciate Khinchine's results we first have to interpret what the width of K in direction $\boldsymbol{d}$ means. (To relate to our topic we assume that K is a full-dimensional polytope P.) To do that it is easier to look at the integer width of P. The integer width of P in the direction $\boldsymbol{d}$ is the number of lattice hyperplanes intersecting P in direction $\boldsymbol{d}$. For a precise definition of a lattice hyperplane we refer to Section 4, but the dashed hyperplanes in Figure 3 give the intuitive explanation. The width $W(\cdot, \cdot)$ is an approximation of the integer width, so Khinchine's result says that if P is lattice point free, then there exists an integer vector $\boldsymbol{d}$ such that the number of lattice hyperplanes intersecting P in direction $\boldsymbol{d}$ is small. The direction $\boldsymbol{d}$ is often referred to as a "thin" direction, and we say that P is "thin" or "flat" in direction $\boldsymbol{d}$. If we could find the direction $\boldsymbol{d}$ in polynomial time (or at least in polynomial time for fixed dimension) then we could solve the integer feasibility problem (1.1) recursively in polynomial time for fixed dimension n. First, we find the direction $\boldsymbol{d}$ and we identify the hyperplanes intersecting P. If P is integer infeasible, then there must be at most a constant number of them if the dimension is fixed. Then, for each of these hyperplanes we consider the polytope formed by the intersection of P with that hyperplane. This is a polytope in dimension less than or equal to $n-1$. For the new polytope we find a new thin direction, and so the process is repeated. We can illustrate the algorithm through a search tree that has at most n levels, and a number of nodes at each level that is bounded by a constant depending only on the dimension on that level.

In Section 4 we describe Lenstra's algorithm, based on the above idea, for solving the integer feasibility problem (1.1) in polynomial time for fixed dimension. In Lenstra's algorithm the flatness result is somehow concealed by a linear transformation, but one of the key components, lattice basis reduction, plays a crucial role in obtaining the integer feasibility algorithm is transparent. What Lenstra does is to apply a linear transformation τ to P so that P becomes more "round". Then, in order not to change the problem, he applies the same transformation to the lattice $\mathbb{Z}^n$. An initial basis for $\mathbb{Z}^n$ is the set of unit vectors $\{\boldsymbol{e}_1, \ldots \boldsymbol{e}_n\}$, and a basis for the transformed lattice $\tau\mathbb{Z}^n$ is the set of vectors $\tau(\boldsymbol{e}_1, \ldots, \boldsymbol{e}_n)$. Using this representation of the lattice can however yield "bad" lattice hyperplanes, so by applying basis reduction to the basis $\tau(\boldsymbol{e}_1, \ldots, \boldsymbol{e}_n)$ we obtain a new so-called reduced basis that gives a good representation. In Section 5 we briefly discuss some related algorithms, and for these algorithms the flatness result is in the foreground and the basis reduction is more concealed.

Before describing Lenstra's algorithm we present some notation, definitions and basic results in Section 2. In Section 3 we present Lovász' basis reduction algorithm, which is a corner stone in Lenstra's algorithm. Finally, in Section 6 we list some open problems.

2. Notation, definitions, and basic results

Let $\boldsymbol{b}_1, \ldots, \boldsymbol{b}_l$ be linearly independent vectors in $\mathbb{R}^n$. The set

$$L = \{\boldsymbol{x} \in \mathbb{R}^n \mid \boldsymbol{x} = \sum_{j=1}^{l} \lambda_j \boldsymbol{b}_j, \lambda_j \in \mathbb{Z},\ 1 \leq j \leq l\} \tag{2.1}$$

is called a *lattice*. The set of vectors $\{\boldsymbol{b}_1, \ldots, \boldsymbol{b}_l\}$ is called a *lattice basis*. The vectors of a lattice L form an additive group, i.e., $\mathbf{0} \in L$, and if $\boldsymbol{x}$ belongs to L, so does $-\boldsymbol{x}$, and if $\boldsymbol{x}, \boldsymbol{y} \in L$, then $\boldsymbol{x} \pm \boldsymbol{y} \in L$. Moreover, the group L is *discrete*, i.e., there exits a real number $r > 0$ such that the n-dimensional ball with radius r, centered at the origin, does not contain any other element from L except the origin.

The *rank* of L, rk L, is equal to the dimension of the Euclidean vector space generated by the basis of L. The rank of the lattice L in expression (2.1) is l, and we have $l \leq n$. If $l = n$ we call the lattice *full-dimensional.*

Let W be the vector space spanned by the lattice L, and let $\boldsymbol{B}_W$ be an orthonormal basis for W. The *determinant of the lattice* L, $d(L)$, is defined as the absolute value of the determinant of any nonsingular linear transformation $W \rightarrow W$ that maps $\boldsymbol{B}_W$ onto a basis of L. Below we give three different formulae for computing $d(L)$. Let $\boldsymbol{B} = (\boldsymbol{b}_1, \ldots, \boldsymbol{b}_l)$ be a basis for the lattice $L \subset \mathbb{R}^n$, with $l \leq n$, and let $\boldsymbol{b}_1^*, \ldots, \boldsymbol{b}_l^*$ be the vectors obtained from applying the Gram-Schmidt orthogonalization procedure to $\boldsymbol{b}_1, \ldots, \boldsymbol{b}_l$ (for more details, see Definition 3.4 in Section 3). Then,

$$d(L) = ||\boldsymbol{b}_1^*|| \cdot ||\boldsymbol{b}_2^*|| \cdot \;\cdots\; \cdot ||\boldsymbol{b}_l^*||,$$

$$d(L) = \sqrt{\det(\boldsymbol{B}^T\boldsymbol{B})}\ ,$$

$$d(L) = \lim_{r \rightarrow \infty} \frac{|\{\boldsymbol{x} \in L : ||\boldsymbol{x}|| < r\}|}{\mathrm{vol}(B_l(r))}, \tag{2.2}$$

where $\mathrm{vol}(B_l(r))$ is the volume of the l-dimensional ball with radius r. If L is full-dimensional, then $d(L)$ can be interpreted as the volume of the parallelepiped $\sum_{j=1}^{n}[0,1)\boldsymbol{b}_j$. In this case the determinant of the lattice can be computed straightforwardly as $d(L) = |\det(\boldsymbol{b}_1, \ldots, \boldsymbol{b}_n)|$. The determinant of $\mathbb{Z}^n$ is equal to one. It is clear from Expression (2.2) that the determinant of a lattice depends only on the lattice and not on the choice of basis, see also Section 3.

For more details about lattices, see e.g. Cassels [**8**], Grötschel, Lovász, and Schrijver [**17**], and Schrijver [**35**].

3. Lovász' basis reduction algorithm

In the lattice approaches to integer programming that will be discussed in Section 4 we need lattice representations using bases with short, almost orthogonal basis vectors. Such bases are called reduced. Here we describe Lovász' basis reduction algorithm [**23**]. We also discuss some recent implementations.

Before discussing specific basis reduction algorithms we describe the basic operations that are used to go from one lattice basis to another.

DEFINITION 3.1. An integer nonsingular matrix $\boldsymbol{U}$ is *unimodular* if $\det(\boldsymbol{U}) = \pm 1$.

The following operations on a matrix are called *elementary column operations*:

- exchanging two columns,
- multiplying a column by -1,
- adding an integer multiple of one column to another column.

THEOREM 3.2. *An integer matrix $\boldsymbol{U}$ is unimodular if and only if $\boldsymbol{U}$ can be derived from the identity matrix by elementary column operations.*

OBSERVATION 3.3. If $\boldsymbol{B}$ and $\boldsymbol{B}'$ are bases for the same lattice $L \subset \mathbb{R}^n$ of rank l, then $\boldsymbol{B}' = \boldsymbol{B}\boldsymbol{U}$ for some $l \times l$ unimodular matrix $\boldsymbol{U}$.

A consequence of Theorem 3.2 and Observation 3.3 is that a lattice of rank 1 has two bases, and a lattice of rank greater than 1 has infinitely many bases. From Observation 3.3 we also see that the determinant of a lattice L only depends on L and not on the chosen basis.

To go from one basis to another is conceptually easy; given a basis $\boldsymbol{B}$ we just multiply $\boldsymbol{B}$ by a unimodular matrix, or equivalently, we perform a series of elementary column operations on $\boldsymbol{B}$, to obtain a new basis. The key question is of course how to do this efficiently so that the new basis is reduced according to the definition of reducedness we are using.

In Lovász' [**23**] basis reduction algorithm the length of the vectors are measured using the Euclidean length, and the Gram-Schmidt vectors corresponding to the current basis are used as a reference for checking whether the basis vectors are nearly orthogonal. Let $L \subset \mathbb{R}^n$ be a lattice, and let $\boldsymbol{b}_1, \ldots, \boldsymbol{b}_l$, $l \le n$, be the current basis vectors for L. We review the Gram-Schmidt process below.

DEFINITION 3.4. The Gram-Schmidt process derives orthogonal vectors $\boldsymbol{b}_j^*$, $1 \le j \le l$, from linearly independent vectors $\boldsymbol{b}_j$, $1 \le j \le l$. The vectors $\boldsymbol{b}_j^*$, $1 \le j \le l$, and the real numbers μ_{jk}, $1 \le k < j \le l$, are determined from $\boldsymbol{b}_j$, $1 \le j \le l$, by the recursion

$$\boldsymbol{b}_1^* = \boldsymbol{b}_1$$

$$\boldsymbol{b}_j^* = \boldsymbol{b}_j - \sum_{k=1}^{j-1} \mu_{jk} \boldsymbol{b}_k^*, \quad 2 \le j \le l,$$

where

$$\mu_{jk} = \frac{\boldsymbol{b}_j^T \boldsymbol{b}_k^*}{||\boldsymbol{b}_k^*||^2}, \quad 1 \le k < j \le l.$$

The vector $\boldsymbol{b}_j^*$ is the projection of $\boldsymbol{b}_j$ on the orthogonal complement of $\sum_{k=1}^{j-1} \mathbb{R}\boldsymbol{b}_k = \{\sum_{k=1}^{j-1} m_k \boldsymbol{b}_k : m_k \in \mathbb{R},\ 1 \le k \le j-1\}$, i.e., $\boldsymbol{b}_j^*$ is the component of $\boldsymbol{b}_j$ orthogonal to the real subspace spanned by $\boldsymbol{b}_1, \ldots, \boldsymbol{b}_{j-1}$. Thus, any pair $\boldsymbol{b}_i^*$, $\boldsymbol{b}_k^*$ of the Gram-Schmidt vectors are mutually orthogonal. The multiplier μ_{jk} gives the length, relative to $\boldsymbol{b}_k^*$, of the component of the vector $\boldsymbol{b}_j$ in direction $\boldsymbol{b}_k^*$. The multiplier μ_{jk} is equal to zero if and only if $\boldsymbol{b}_j$ is orthogonal to $\boldsymbol{b}_k^*$. Notice that the Gram-Schmidt vectors corresponding to $\boldsymbol{b}_1, \ldots, \boldsymbol{b}_l$ do not necessarily belong to L, but they do span the same real vector space as $\boldsymbol{b}_1, \ldots, \boldsymbol{b}_l$.

DEFINITION 3.5 ([**23**]). A basis $\boldsymbol{b}_1, \boldsymbol{b}_2, \ldots, \boldsymbol{b}_l$ is called *reduced* if

$$|\mu_{jk}| \leq \frac{1}{2} \quad \text{for } 1 \leq k < j \leq l, \tag{3.1}$$

$$||\boldsymbol{b}_j^* + \mu_{j,j-1}\boldsymbol{b}_{j-1}^*||^2 \geq \frac{3}{4}||\boldsymbol{b}_{j-1}^*||^2 \quad \text{for } 1 < j \leq l. \tag{3.2}$$

Below we explain why vectors satisfying Conditions (3.1) and (3.2) are relatively short and nearly orthogonal.

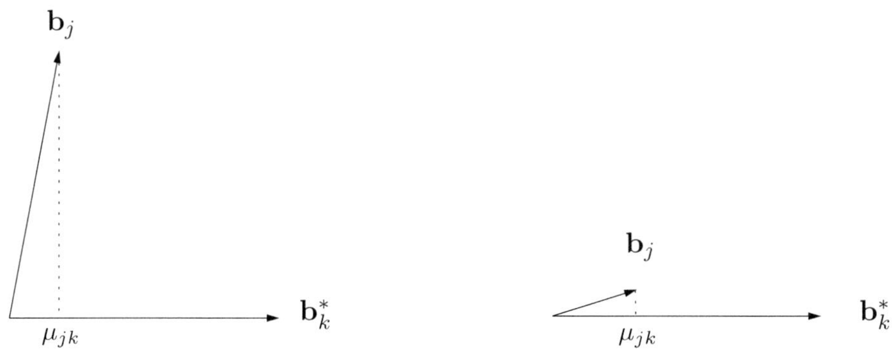

(a) $\mu_{jk} < 1/2$, since $\mathbf{b}_j$ is almost orthogonal to $\mathbf{b}_k^*$.

(b) $\mu_{jk} < 1/2$ since $\mathbf{b}_j$ is short compared to $\mathbf{b}_k^*$. If $k = j-1$, Condition (9) is violated.

FIGURE 4. Cases for which Condition (3.1) are satisfied.

Condition (3.1) is satisfied in two cases. The first case (see Figure 4(a),) is if $\boldsymbol{b}_j$ is almost orthogonal to $\boldsymbol{b}_k^*$. Then, clearly, if we project $\boldsymbol{b}_j$ on $\boldsymbol{b}_k^*$, the absolute value of the length of this projection is going to be short relative to the length of $\boldsymbol{b}_k^*$. The second possibility for (3.1) to be satisfied (see Figure 4(b)) is if $\boldsymbol{b}_j$ is short relative to $\boldsymbol{b}_k^*$. Even if $\boldsymbol{b}_j$ and $\boldsymbol{b}_k^*$ are not close to being orthogonal, the length of the projection of $\boldsymbol{b}_j$ on $\boldsymbol{b}_k^*$ will still be small relative to the length of $\boldsymbol{b}_k^*$. If we would accept this case we would also accept a basis in which $||\boldsymbol{b}_1|| >> ||\boldsymbol{b}_2|| >> \cdots >> ||\boldsymbol{b}_l||$, and where the vectors are far from being orthogonal. To prevent this Condition (3.2) is enforced. Here we relate to the interpretation of the Gram-Schmidt vectors above, and notice that the vectors $\boldsymbol{b}_j^* + \mu_{j,j-1}\boldsymbol{b}_{j-1}^*$ and $\boldsymbol{b}_{j-1}^*$ are the projections of $\boldsymbol{b}_j$ and $\boldsymbol{b}_{j-1}$ on the orthogonal complement of $\sum_{k=1}^{j-2} \mathbb{R}\boldsymbol{b}_k$. Consider a violation of Condition (3.1) where $k = j-1$, i.e., suppose that $\boldsymbol{b}_j$ is short compared to $\boldsymbol{b}_{j-1}^*$, which implies that $\boldsymbol{b}_j^*$ is short compared to $\boldsymbol{b}_{j-1}^*$ as $||\boldsymbol{b}_j^*|| \leq ||\boldsymbol{b}_j||$. Suppose we *interchange* $\boldsymbol{b}_j$ and $\boldsymbol{b}_{j-1}$. Then the new $\boldsymbol{b}_{j-1}^*$ will be the old $\boldsymbol{b}_j^* + \mu_{j,j-1}\boldsymbol{b}_{j-1}^*$, which will be short compared to the old $\boldsymbol{b}_{j-1}^*$, i.e., Condition (3.2) will be violated.

If Condition (3.1) is violated, i.e., the component of vector $\boldsymbol{b}_j$ in direction $\boldsymbol{b}_k^*$ is *relatively long*, then Lovász' basis reduction algorithm will replace $\boldsymbol{b}_j$ by $\boldsymbol{b}_j - \lceil \mu_{jk} \rfloor \boldsymbol{b}_k$, where $\lceil \mu_{jk} \rfloor := \lceil \mu_{jk} - \frac{1}{2} \rceil$. Such a step is called *size reduction* and will ensure relatively short basis vectors. If Condition (3.2) is violated for a certain index j, then the vectors $\boldsymbol{b}_j$ and $\boldsymbol{b}_{j-1}$ are *interchanged* to prevent us from accepting a basis with long non-orthogonal vectors as described in the previous paragraph.

It is clear that both the size reduction and the interchange are elementary column operations.

To summarize, Lovász' algorithm applies a sequence of *size reductions* and *interchanges* in order to reduce the length of the vectors, and to prevent us from obtaining non-orthogonal basis vectors of decreasing length, where the first basis vector may be arbitrarily long. The constant $\frac{3}{4}$ in inequality (3.2) is arbitrarily chosen and can be replaced by any fixed real number $\frac{1}{4} < y < 1$. In a practical implementation one chooses a constant close to one.

A brief outline of Lovász' basis reduction algorithm is as follows. For precise details we refer to [**23**]. First compute the Gram-Schmidt vectors $\boldsymbol{b}_j^*$, $1 \le j \le l$, and the numbers μ_{jk}, $1 \le k < j \le l$. Initialize $i := 2$. Perform, if necessary, a size reduction to obtain $|\mu_{i,i-1}| \le 1/2$. Update $\mu_{i,i-1}$. Then check whether Condition (3.2) holds for $j = i$. If Condition (3.2) is violated, then exchange $\boldsymbol{b}_i$ and $\boldsymbol{b}_{i-1}$, and update the relevant Gram-Schmidt vectors and numbers μ_{jk}. If $i > 2$, then let $i := i - 1$. Next, achieve $|\mu_{im}| \le 1/2$ for $m = i-2, i-3, \ldots, 1$. If $i = l$, stop. Otherwise, let $i := i + 1$.

The key to the termination argument of Lovász' algorithm is the following potential function $\Phi(\boldsymbol{b}_1, \ldots, \boldsymbol{b}_n)$ of a lattice basis $\boldsymbol{B} = (\boldsymbol{b}_1, \ldots, \boldsymbol{b}_n)$, $\boldsymbol{b}_j \in \mathbb{Z}^n$, $1 \le j \le n$,

$$\Phi(\boldsymbol{B}) = \|\boldsymbol{b}_1^*\|^{2n} \|\boldsymbol{b}_2^*\|^{2(n-1)} \cdots \|\boldsymbol{b}_n^*\|^2.$$

The potential of an integer lattice basis is always an integer. Furthermore, an *interchange* step in Lovász' algorithm decreases the potential by a factor of 3/4 or a smaller number. Thus, if $\boldsymbol{B}_1$ and $\boldsymbol{B}_2$ are two subsequent bases after an interchange step in Lovász' algorithm, then

$$\Phi(\boldsymbol{B}_2) \le \frac{3}{4}\Phi(\boldsymbol{B}_1).$$

The potential of the input basis $\boldsymbol{B}$ can be bounded by $\Phi(\boldsymbol{B}) \le (\|\boldsymbol{b}_1\| \cdots \|\boldsymbol{b}_n\|)^{2n}$. The number of iterations can thus be bounded by $\mathrm{O}(n(\log \|\boldsymbol{b}_1\| + \ldots + \log \|\boldsymbol{b}_n\|))$. Taking the updating steps into account and with an analysis of the size of the numbers produced in the course of the algorithm, one obtains the following bound on the running time.

THEOREM 3.6 ([**23**]). *Let $L \subseteq \mathbb{Z}^n$ be a lattice with basis $\boldsymbol{b}_1, \ldots, \boldsymbol{b}_n$, and let $\beta \in \mathbb{R}$, $\beta \ge 2$, be such that $\|\boldsymbol{b}_j\|^2 \le \beta$ for $1 \le j \le n$. Then the number of arithmetic operations needed by the basis reduction algorithm as described in* [**23**] *is $O(n^4 \log \beta)$, and the integers on which these operations are performed each have binary length $O(n \log \beta)$.*

In terms of bit operations, Theorem 3.6 implies that Lovász' basis reduction algorithm has a running time of $O(n^6 (\log \beta)^3)$ using classical algorithms for addition and multiplication.

EXAMPLE 3.7. Here we give an example of an initial and a reduced basis for a given lattice. Let L be the lattice generated by the vectors

$$\boldsymbol{b}_1 = \begin{pmatrix} 4 \\ 1 \end{pmatrix} \quad \boldsymbol{b}_2 = \begin{pmatrix} 1 \\ 1 \end{pmatrix}.$$

The Gram-Schmidt vectors are $\boldsymbol{b}_1^* = \boldsymbol{b}_1$ and $\boldsymbol{b}_2^* = \boldsymbol{b}_2 - \mu_{21}\boldsymbol{b}_1^* = (1,1)^T - \frac{5}{17}\boldsymbol{b}_1^* = \frac{1}{17}(-3, 12)^T$, see Figure 5a. Condition (3.1) is satisfied since $\boldsymbol{b}_2$ is short relative to

b_1^*. However, Condition (3.2) is violated, so we exchange b_1 and b_2, giving

$$b_1 = \begin{pmatrix} 1 \\ 1 \end{pmatrix} \quad b_2 = \begin{pmatrix} 4 \\ 1 \end{pmatrix}.$$

We now have $b_1^* = b_1$, $\mu_{21} = \frac{5}{2}$ and $b_2^* = \frac{1}{2}(3, -3)^T$, see Figure 5b.

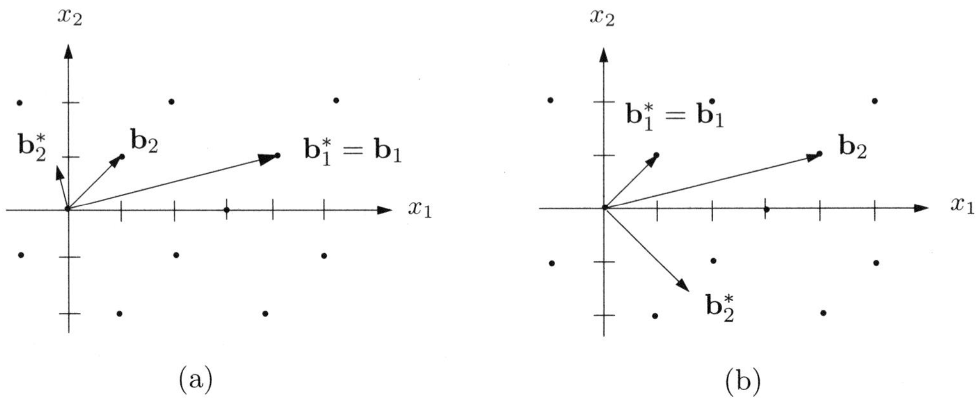

FIGURE 5

Condition (3.1) is now violated, so we replace b_2 by $b_2 - 2b_1 = (2, -1)^T$. Conditions (3.1) and (3.2) are satisfied for the resulting basis

$$b_1 = \begin{pmatrix} 1 \\ 1 \end{pmatrix} \quad b_2 = \begin{pmatrix} 2 \\ -1 \end{pmatrix},$$

and hence this basis is reduced, see Figure 6.

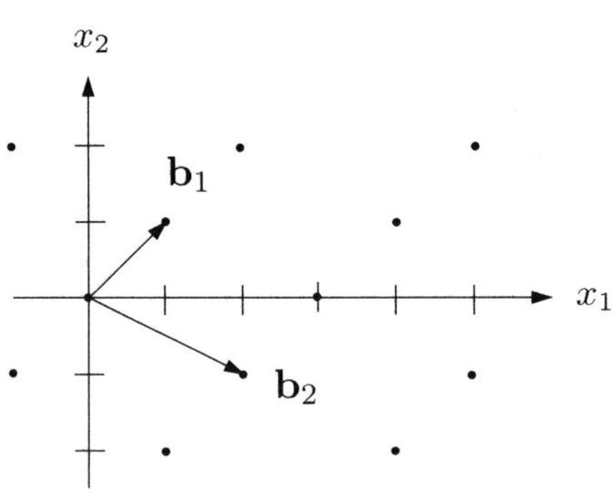

FIGURE 6. The reduced basis.

□

Next we will present some useful bounds on reduced basis vectors. In Propositions 3.8 and 3.10 we assume that the lattice L is full-dimensional.

PROPOSITION 3.8 ([**23**]). *Let $b_1, \ldots, b_n$ be a reduced basis for the lattice $L \subset \mathbb{R}^n$. Then,*

$$d(L) \le \Pi_{j=1}^n ||b_j|| \le c_1 \cdot d(L), \tag{3.3}$$

where $c_1 = 2^{n(n-1)/4}$.

The first inequality in (3.3) is also referred to as the *inequality of Hadamard* that holds for any basis of L. Hadamard's inequality holds with equality if and only if the basis is orthogonal. The ratio $\Pi_{j=1}^{n}||\boldsymbol{b}_j||/d(L)$ is called the *orthogonality defect* of the basis. Hermite [**18**] proved that each lattice $L \subset \mathbb{R}^n$ has a basis $\boldsymbol{b}_1, \ldots, \boldsymbol{b}_n$ such that $\Pi_{j=1}^{n}||\boldsymbol{b}_j|| \leq c \cdot d(L)$, where c is a constant depending only on n. The basis produced by Lovász' basis reduction algorithm yields the constant $c = c_1$ in Proposition 3.8. Better constants than c_1 are possible, but the question is then whether the basis can be obtained in polynomial time.

A consequence of Proposition 3.8 is that if we consider a basis that satisfies (3.3), then the distance of the basis vector $\boldsymbol{b}_n$ to the hyperplane generated by the reduced basis vectors $\boldsymbol{b}_1, \ldots, \boldsymbol{b}_{n-1}$ is not too small as stated in the following corollary.

COROLLARY 3.9 ([**25**]). *Assume that* $\boldsymbol{b}_1, \ldots, \boldsymbol{b}_n$ *is a basis such that (3.3) holds, and that, after possible reordering,* $||\boldsymbol{b}_n|| = \max_{1\leq j\leq n}\{||\boldsymbol{b}_j||\}$. *Let* $H = \sum_{j=1}^{n-1}\mathbb{R}\boldsymbol{b}_j$ *and let* h *be the distance of the basis vector* $\boldsymbol{b}_n$ *to* H. *Then*

$$c_1^{-1} \cdot ||\boldsymbol{b}_n|| \leq h \leq ||\boldsymbol{b}_n||, \tag{3.4}$$

where $c_1 = 2^{n(n-1)/4}$.

The lower bound on h given in Corollary 3.9 plays a crucial role in the algorithm of H. W. Lenstra, Jr., that is described in Section 4.

PROPOSITION 3.10 ([**23**]). *Let* $L \subset \mathbb{R}^n$ *be a lattice with reduced basis* $\boldsymbol{b}_1, \ldots, \boldsymbol{b}_n \in \mathbb{R}^n$. *Let* $\boldsymbol{x}_1, \ldots, \boldsymbol{x}_t \in L$ *be linearly independent. Then we have*

$$||\boldsymbol{b}_1||^2 \leq 2^{n-1}||\boldsymbol{x}||^2 \quad \text{for all } \boldsymbol{x} \in L,\ \boldsymbol{x} \neq 0, \tag{3.5}$$

$$||\boldsymbol{b}_j||^2 \leq 2^{n-1}\max\{||\boldsymbol{x}_1||^2, ||\boldsymbol{x}_2||^2, \ldots, ||\boldsymbol{x}_t||^2\} \quad \text{for } 1 \leq j \leq t. \tag{3.6}$$

Inequality (3.5) implies that the first reduced basis vector $\boldsymbol{b}_1$ is an approximation of the shortest nonzero vector in L.

Just as the first basis vector is an approximation of the shortest vector of the lattice (3.5), the other basis vectors are approximations of the *successive minima* of the lattice. The j-th successive minimum of $||\ ||$ on L is the smallest positive value ν_j such that there exists j linearly independent elements of the lattice L in the ball of radius ν_j centered at the origin.

PROPOSITION 3.11 ([**23**]). *Let* $\nu_1, \ldots, \nu_l$ *denote the successive minima of* $||\ ||$ *on* L, *and let* $\boldsymbol{b}_1, \ldots, \boldsymbol{b}_l$ *be a reduced basis for* L. *Then*

$$2^{(1-j)/2}\nu_j \leq ||\boldsymbol{b}_j|| \leq 2^{(l-1)/2}\nu_j \quad \text{for } 1 \leq j \leq l.$$

In recent years several new variants of Lovász' basis reduction algorithm have been developed, and a number of variants for implementation have been suggested. We mention a few below and recommend the paper by Schnorr and Euchner [**33**] for a more detailed overview. Schnorr [**31**] extended Lovász' algorithm to a family of polynomial time algorithms that, given $\varepsilon > 0$, finds a non-zero vector in an n-dimensional lattice that is no longer than $(1+\varepsilon)^n$ times the length of the shortest vector in the lattice. The degree of the polynomial that bounds the running time of the family of algorithms increases as ε goes to zero. Seysen [**36**] developed an algorithm in which the intermediate integers that are produced are no larger than the input integers. Seysen's algorithm performs well particularly on lower-dimensional lattices. Schnorr and Euchner [**33**] discuss the possibility of computing the Gram-Schmidt vectors using floating point arithmetic while keeping the basis vectors in exact arithmetic in order to improve the practical performance of the algorithm. The drawback of this approach is that the basis reduction algorithm might become unstable. They propose a floating point version with good stability, but cannot prove that the algorithm always terminates. Their computational study indicates that their version is stable on instances of dimension up to 125 having input numbers of bit length as large as 300. Our experience is that one can use basis reduction for problems of larger dimensions if the input numbers are smaller, but once the dimension reaches about 300-400, basis reduction will be slow. Another version considered by Schnorr and Euchner is basis reduction *with deep insertions*. Here, they allow for a vector $\boldsymbol{b}_j$ to be swapped with a vector with lower index than $j-1$. Schnorr [**31, 32**] also developed a variant of Lovász' algorithm in which not only two vectors are interchanged during the reduction process, but where blocks $\boldsymbol{b}_j, \boldsymbol{b}_{j+1}, \ldots, \boldsymbol{b}_{j+\beta-1}$ of β consecutive vectors are transformed so as to minimize the j-th Gram Schmidt vector $\boldsymbol{b}_j^*$. This so called *block reduction* produces shorter basis vectors but needs more computing time. The shortest vector $\boldsymbol{b}_j^*$ in a block of size β is determined by complete enumeration of all short lattice vectors. Schnorr and Hörner [**34**] develop and analyze a rule for pruning this enumeration process.

For the reader interested in using a version of Lovász' basis reduction algorithm there are some useful libraries available on the Internet. Two of them are LiDIA – a C++ Library for Computational Number Theory [**26**], and NTL – a Library for doing Number Theory, developed by V. Shoup [**37**].

4. Lenstra's algorithm

As we saw in Example 1.1, if one uses branch-and-bound for solving problem (1.1) it is possible, even in dimension 2, to create an arbitrarily deep search tree for certain thin polytopes. Lenstra [**25**] suggested transforming the polytope P using a linear transformation τ such that the polytope τP becomes "round" according to a certain measure. Assume without loss of generality that the polytope P is full-dimensional and bounded, and let $B(\boldsymbol{p}, z) = \{\boldsymbol{x} \in \mathbb{R}^n : ||\boldsymbol{x} - \boldsymbol{p}|| \leq z\}$ be the closed ball with center $\boldsymbol{p}$ and radius z. The transformation τ that we apply to the polytope is constructed so that $B(\boldsymbol{p}, r) \subset \tau P \subset B(\boldsymbol{p}, R)$ for some $\boldsymbol{p} \in \tau P$, with r and R satisfying

$$\frac{R}{r} \leq c_2, \tag{4.1}$$

where c_2 is a constant that depends only on the dimension n. Relation (4.1) is the measure of "roundness" that Lenstra uses. For an illustration, see Figure 7. Once we have transformed the polytope, we need to apply the same transformation to the lattice, which gives us the following feasibility problem that is equivalent to problem (1.1):

$$\text{Is } \tau\mathbb{Z}^n \cap \tau P \neq \emptyset? \tag{4.2}$$

The vectors $\tau \boldsymbol{e}_j$, $1 \leq j \leq n$, where $\boldsymbol{e}_j$ is the j-th unit vector in $\mathbb{R}^n$, form a basis for the lattice $\tau\mathbb{Z}^n$. If the polytope P is thin, then this will translate to the lattice basis vectors $\tau \boldsymbol{e}_j$, $1 \leq j \leq n$, in the sense that these vectors are long and non-orthogonal.

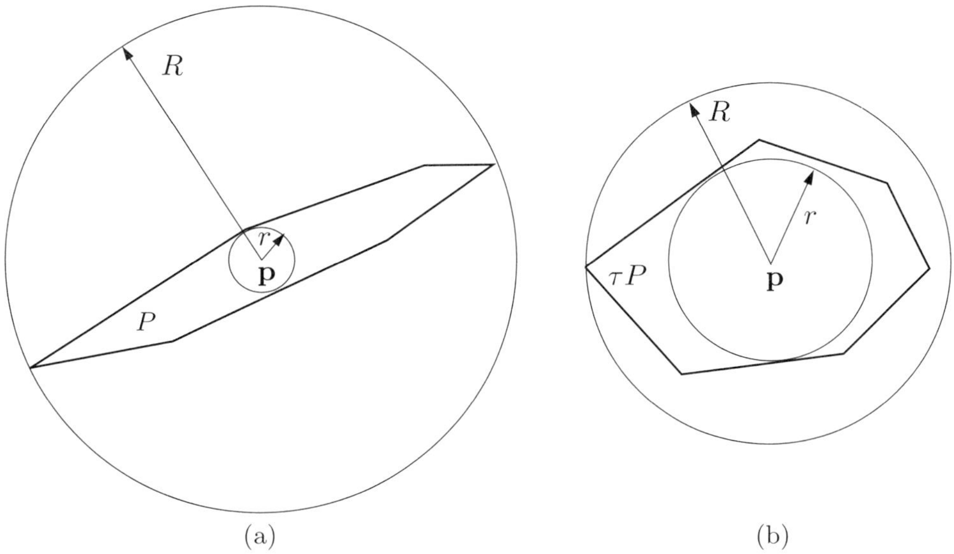

FIGURE 7. (a) The original polytope P is thin, and the ratio R/r is large.
(b) The transformed polytope τP is "round", and R/r is relatively small.

This is where lattice basis reduction becomes useful. Once we have the transformed polytope τP, Lenstra uses the following lemma to find a lattice point quickly.

LEMMA 4.1 ([**25**]). *Let $\boldsymbol{b}_1, \ldots, \boldsymbol{b}_n$ be any basis for L. Then for all $\boldsymbol{x} \in \mathbb{R}^n$ there exists a vector $\boldsymbol{y} \in L$ such that*

$$||\boldsymbol{x} - \boldsymbol{y}||^2 \leq \frac{1}{4}(||\boldsymbol{b}_1||^2 + \cdots + ||\boldsymbol{b}_n||^2).$$

The proof of this lemma suggests a fast construction of the vector $\boldsymbol{y} \in L$ given the vector $\boldsymbol{x}$.

Next, let $L = \tau\mathbb{Z}^n$, and let $\boldsymbol{b}_1, \ldots, \boldsymbol{b}_n$ be a basis for L such that (3.3) holds. Notice that (3.3) holds if the basis is reduced. Also, reorder the vectors so that $||\boldsymbol{b}_n|| = \max_{1 \leq j \leq n}\{||\boldsymbol{b}_j||\}$. After reordering, the basis might not be reduced any longer, but Condition (3.3) still holds. Let $\boldsymbol{x} = \boldsymbol{p}$ where $\boldsymbol{p}$ is the center of the closed balls $B(\boldsymbol{p}, r)$ and $B(\boldsymbol{p}, R)$. Apply Lemma 4.1 to the given $\boldsymbol{x}$. This gives a lattice vector $\boldsymbol{y} \in \tau\mathbb{Z}^n$ such that

$$||\boldsymbol{p} - \boldsymbol{y}||^2 \leq \frac{1}{4}(||\boldsymbol{b}_1||^2 + \cdots + ||\boldsymbol{b}_n||^2) \leq \frac{1}{4} \cdot n \cdot ||\boldsymbol{b}_n||^2 \tag{4.3}$$

in polynomial time. We now distinguish two cases. Either $\boldsymbol{y} \in \tau P$ or $\boldsymbol{y} \notin \tau P$. In the first case we are done, so assume we are in the second case. Since $\boldsymbol{y} \notin \tau P$ we know that $\boldsymbol{y}$ is not inside the ball $B(\boldsymbol{p}, r)$ as $B(\boldsymbol{p}, r)$ is completely contained in τP. Hence we know that $||\boldsymbol{p} - \boldsymbol{y}|| > r$, or using (4.3), that

$$r < \frac{1}{2} \cdot \sqrt{n} \cdot ||\boldsymbol{b}_n||. \tag{4.4}$$

Below we will describe the tree search algorithm and argue why it is polynomial for fixed n. The distance between any two consecutive hyperplanes is equal to a certain constant. We now create t subproblems by considering intersections between the polytope τP with t of these parallel hyperplanes. Each of the subproblems has dimension at least one lower than the parent problem, and they are solved recursively. The procedure of splitting the problem into subproblems of lower dimension is called "branching", and each subproblem is represented by a node in the enumeration tree. In each node we repeat the whole process of transformation, basis reduction and, if necessary, branching. The enumeration tree created by this recursive process is of depth at most n, and the number of nodes t at each level is bounded by a constant that depends only on the dimension. The value of t will be computed below.

Let H and h be defined as in Corollary 3.9 of Section 3, and let $L' = \sum_{j=1}^{n-1} \mathbb{Z}\boldsymbol{b}_j$. We can write L as

$$L = L' + \mathbb{Z}\boldsymbol{b}_n \subset H + \mathbb{Z}\boldsymbol{b}_n = \cup_{k\in\mathbb{Z}}(H + k\boldsymbol{b}_n). \tag{4.5}$$

So the lattice L is contained in countably many parallel hyperplanes. For an example we refer to Figure 8. The distance between two consecutive hyperplanes is

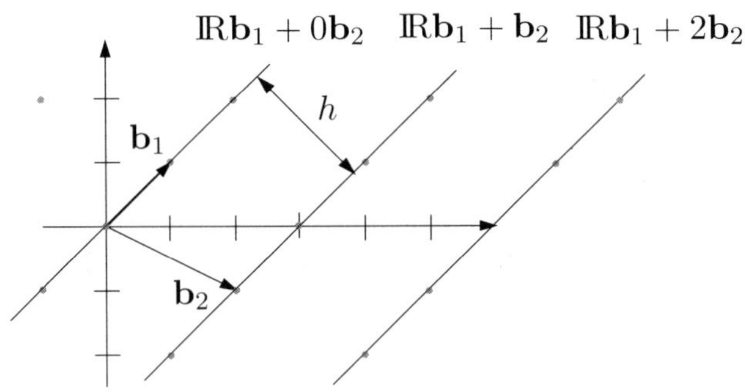

FIGURE 8

h, and Corollary 3.9 says that h is bounded from below by $c_1^{-1}||\boldsymbol{b}_n||$, which implies that not too many hyperplanes intersect τP. To determine precisely how many hyperplanes intersect τP, we approximate τP by the ball $B(\boldsymbol{p}, R)$. If t is the number of hyperplanes intersecting $B(\boldsymbol{p}, R)$ we have

$$t - 1 \le \frac{2R}{h}.$$

Using the relationship (4.1) between the radii R and r we have $2R \le 2rc_2 < c_2\sqrt{n}||\boldsymbol{b}_n||$, where the last inequality follows from (4.4). Since $h \ge c_1^{-1}||\boldsymbol{b}_n||$, we get the following bound on the number of hyperplanes that we need to consider:

$$t - 1 \le \frac{2R}{h} < c_1c_2\sqrt{n},$$

which depends on the dimension only. The values of the constants c_1 and c_2 that are used by Lenstra are: $c_1 = 2^{n(n-1)/4}$ and $c_2 = 2n^{3/2}$. Lenstra discusses ways of improving these values. To determine the values of k in expression (4.5), we express $\boldsymbol{p}$ as a linear combination of the basis vectors $\boldsymbol{b}_1, \ldots, \boldsymbol{b}_n$. Recall that $\boldsymbol{p}$ is the center of the ball $B(\boldsymbol{p}, R)$ that was used to approximate τP.

So far we have not mentioned how to determine the transformation τ and hence the balls $B(\boldsymbol{p}, r)$ and $B(\boldsymbol{p}, R)$. We give the general idea here without going into detail. First, determine an n-simplex contained in P. This can be done by repeated calls to the ellipsoid algorithm. The resulting simplex is described by its extreme points $\boldsymbol{v}_0, \ldots, \boldsymbol{v}_n$. By applying the ellipsoid algorithm repeatedly we can decide whether there exists an extreme point $\boldsymbol{x}$ of P such that if we replace $\boldsymbol{v}_j$ by $\boldsymbol{x}$ we obtain a new simplex whose volume is at least a factor of $\frac{3}{2}$ larger than the current simplex. We stop the procedure if we cannot find such a new simplex. The factor $\frac{3}{2}$ can be modified, but the choice will affect the value of the constant c_2, see [**25**] for further details. We now map the extreme points of the simplex to the unit vectors of $\mathbb{R}^{n+1}$ so as to obtain a regular n-simplex with rational coordinates, and we denote this transformation by τ. Lenstra [**25**] shows that τ has the property that if we let $\boldsymbol{p} = 1/(n+1)\sum_{j=0}^{n} \boldsymbol{e}_j$, where $\boldsymbol{e}_j$ is the j-th unit vector of $\mathbb{R}^{n+1}$ (i.e., $\boldsymbol{p}$ is the center of the regular simplex), then there exist closed balls $B(\boldsymbol{p}, r)$ and $B(\boldsymbol{p}, R)$ such that $B(\boldsymbol{p}, r) \subset \tau P \subset B(\boldsymbol{p}, R)$ and such that $R/r \leq c_2$.

Lenstra's algorithm has been implemented by Gao and Zhang [**15**], and a heuristic version of the algorithm has been developed and implemented by Aardal et al. [**1**], and Aardal and Lenstra [**4**]

5. A few related results

In the working paper by Lenstra [**24**], he used a basis reduction algorithm that is polynomial for fixed n only. Later Lovász developed the basis reduction algorithm described in Section 3. This is the basis reduction algorithm used in the published paper by Lenstra [**25**], so even though the result in Theorem 5.1 below is present in Lenstra's paper, the result is credited to Lovász (see for instance Schrijver [**35**], pp 256–259). Notice that "polynomial time" in Theorem 5.1 is polynomial time for *varying* n.

THEOREM 5.1. *Let $\boldsymbol{Ax} \leq \boldsymbol{b}$ be a system of m rational inequalities in n variables, let $P = \{\boldsymbol{x} \in \mathbb{R}^n \mid \boldsymbol{Ax} \leq \boldsymbol{b}\}$, and let $W(P, \boldsymbol{d})$ be defined as in Definition 1.3. There exists a polynomial time algorithm that finds either an integer vector $\boldsymbol{y} \in P$, or a vector $\boldsymbol{d} \in \mathbb{Z}^n \setminus \{\boldsymbol{0}\}$ such that*

$$W(P, \boldsymbol{d}) \leq 2n(n+1)2^{n(n-1)/4} \tag{5.1}$$

Lenstra's result that Problem 1.1 can be solved in polynomial time for fixed number of variables can be obtained as a corollary to Theorem 5.1 in the following way. If we apply the algorithm implied by Theorem 5.1, we either find an integer point $\boldsymbol{y} \in P$ or a thin direction $\boldsymbol{d}$, i.e., a direction $\boldsymbol{d}$ such that inequality (5.1) holds. Assume that the direction $\boldsymbol{d}$ is the outcome of the algorithm. Let $\mu = \lceil \min\{\boldsymbol{dx} \mid \boldsymbol{x} \in P\} \rceil$. All points in $P \cap \mathbb{Z}^n$ are contained in the parallel hyperplanes $\boldsymbol{dx} = t$ where $t = \mu, \ldots, \mu + 2n(n+1)2^{n(n-1)/4}$, so if n is fixed, the number of hyperplanes intersecting P is bounded by a constant depending only on n. Each of these hyperplanes gives rise to a subproblem of dimension less than or equal to $n-1$. For each of these lower-dimensional problems we repeat the algorithm of Theorem 5.1. The search tree has at most n levels and the number of nodes at each level is bounded by a constant debpending on the dimension at that level.

Kannan [**20**] developed a variant of Lenstra's algorithm. The algorithm follows Lenstra's algorithm up to the point where he has applied a linear transformation to the polytope P and obtained a polytope τP such that $B(\boldsymbol{p}, r) \subset \tau P \subset B(\boldsymbol{p}, R)$ for some $\boldsymbol{p} \in \tau P$. Here Kannan proceeds as follows. He applies a reduction algorithm to a basis of the lattice $\tau\mathbb{Z}^n$ that produces a "reduced" basis defined differently from a Lovász' reduced basis. In particular, in Kannan's reduced basis the first basis vector is the shortest nonzero lattice vector. As in Lenstra's algorithm two cases are considered. Either τP is relatively large, which implies that τP contains a lattice vector, or τP is small, which means that not too many lattice hyperplanes can intersect τP. Each such intersection gives rise to a subproblem of at least one dimension lower. Kannan's reduced basis makes it possible to improve the bound on the number of hyperplanes that has to be considered to $O(n^{5/2})$.

Lovász and Scarf [**27**] developed a basis reduction algorithm called "generalized basis reduction". Here, the norm used is related to the polytope P. Based on their basis reduction algorithm, and several other results, they suggested an integer programming algorithm that also uses the "branching on hyperplanes" idea, where we branch in "thin" directions. A heuristic version of Lovász and Scarf's integer programming algorithm has been implemented by Cook et al. [**10**].

Barvinok [**6**] generalized Lenstra's result by giving an algorithm that counts integer points in a polytope in polynomial time if the dimension is fixed. Later, Dyer and Kannan [**13**] developed a simplification of Barvinok's algorithm. Barvinok's algorithm has been implemented by De Loera et al. [**12**] who also introduced further practical improvements.

Cook, Hartmann, Kannan, and McDiarmid [**9**] proved that if P_I is the integer hull of the rational polyhedron $P \subset \mathbb{R}^n$ given by m inequalities whose size is at most φ, then for fixed n an upper bound on the number of vertices of P_I is $\mathrm{O}(m^n\varphi^{n-1})$. This implies that integer hull computations can be done in polynomial time for fixed n as well.

The integer optimization problem $\max\{\boldsymbol{cx} \mid \boldsymbol{Ax} \leq \boldsymbol{d}\}$ can be reduced to the integer feasibility problem by applying binary search. The running time of such an algorithm is $\mathrm{O}(m\varphi + \varphi^2)$, where φ is the maximum size of any of the constraints and the objective vector $\boldsymbol{c}$. Eisenbrand [**14**] developed an algorithm for the integer optimization problem with a fixed number, m, of constraints that uses $\mathrm{O}(\varphi)$ arithmetic operations on rational numbers of binary encoding length $\mathrm{O}(\varphi)$. By combining his results with a randomized algorithm by Clarkson [**11**], Eisenbrand obtains an $\mathrm{O}(m + \log m\,\varphi)$ algorithm for the integer optimization problem.

6. Some open problems and further reading

Computationally there is a lot to be done with respect to branching on hyperplane algorithms. The few implementations that exist (see e.g. [**1, 4, 10, 15**]) all seem to agree that fewer, in many cases significantly fewer, search nodes are needed compared to branch-and-bound. Each node, however, requires more work as a good search direction has to be computed instead of just solving a linear programming relaxation. More work is needed on determining "practically good" search directions, rather than just more or less following the theoretical proofs of the various algorithms, in order to profit from the gain in the reduced size of the search trees. Another bottleneck is the basis reduction algorithm. While polynomial, it still becomes slow when the problem size exceeds a couple of hundred variables. Would it be possible to develop a faster basis reduction algorithm with no worse worst case bounds than the ones given in Propositions 3.10 and 3.11? Goldreich and Goldwasser [**16**] have shown that approximating the shortest nonzero vector in a given lattice with an approximation bound of $O(\sqrt{n})$ is not NP-hard unless the polynomial hierarchy collapses. It should be pointed out though that, in practice, Lovász' basis reduction algorithm produces basis vectors that are much better than the upper bounds in Propositions 3.10 and 3.11 suggest. This is one of the reasons why the algorithm has become such a useful tool in many applications such as cryptography, see eg. Nguyen and Stern [**30**]. Why the algorithm performs so well is not well understood.

For further reading on this topic we refer to the overview article by Kannan [**19**], and the books by Grötschel, Lovász, and Schrijver, [**17**], and by Schrijver[**35**]. Branch-and-bound and cutting plane related integer programming algorithms are treated in depth in the book by Nemhauser and Wolsey [**29**]. Aardal and Eisenbrand [**2**] give an overview on lattice based integer programming as well as other results on integer programming in fixed dimension.

Acknowledgements

We wish to thank Jesus De Loera and two anonymous reviewers for the many insightful comments.

References

[1] K. Aardal, R. E. Bixby, C. A. J. Hurkens, A. K. Lenstra, and J. W. Smeltink, *Market split and basis reduction: towards a solution of the Cornuéjols-Dawande instances*, INFORMS Journal on Computing **12(3)** (2000), 192–202.

[2] K. Aardal, F. Eisenbrand, *Integer programming, lattices, and results in fixed dimension.* To appear as a chapter in "Handbook on Discrete Optimization" in the series "Handbooks in Operations research and Management Science", Elsevier, Amsterdam.

[3] K. Aardal, C. A. J. Hurkens, and A. K. Lenstra, *Solving a system of linear Diophantine equations with lower and upper bounds on the variables*, Mathematics of Operations Research **25(3)** (2000), 427–442.

[4] K. Aardal and A. K. Lenstra, *Hard equality constrained integer knapsacks*, to appear in Mathematics of Operations Research.

[5] W. Banaszczyk, A. E. Litvak, A. Pajor, and S. J. Szarek, *The flatness theorem for non-symmetric convex bodies via the local theory of Banach spaces*, Mathematics of Operations Research, **24(3)**(1999), 728–750.

[6] A. I. Barvinok, *A polynomial time algorithm for counting integral points in polyhedra when the dimension is fixed*, Mathematics of Operations Research **19(4)** (1994), 769–779.

[7] I. Borosh and L. B. Treybig, *Bounds on positive integral solutions of linear diophantine equations*, Proceedings of the American Mathematical Society, vol. 55, American Mathematical Society, Providence, RI, 1976, pp. 299–304.

[8] J. W. S. Cassels, *An Introduction to the Geometry of Numbers.* Second Printing, Corrected. Reprint of the 1971 ed. Springer-Verlag, Berlin, Heidelberg, 1997.

[9] W. Cook, M. E. Hartmann, R. Kannan, and C. McDiarmid, *On integer points in polyhedra*, Combinatorica **12(1)** (1992), 27–37.

[10] W. Cook, T. Rutherford, H. E. Scarf, and D. Shallcross. *An implementation of the generalized basis reduction algorithm for integer programming*, ORSA Journal on Computing **5(2)** (1993) 206–212.

[11] K. L. Clarkson. *Las Vegas algorithms for linear and integer programming when the dimension is small*, Journal of the Association for Computing Machinery **42** (1995), 448–499.

[12] J. A. De Loera, R. Hemmecke, J. Tauzer, and R. Yoshida, *Effective lattice point counting in rational polytopes*, 2002, to appear in Journal of Symbolic Computation. Available at: `http://www.math.ucdavis.edu/~deloera`.

[13] M. E. Dyer and R. Kannan, *On Barvinok's algorithm for counting lattice points in fixed dimension*, Mathematics of Operations Research **22(3)** (1997), 545–549.

[14] F. Eisenbrand, *Fast integer programming in fixed dimension*, in: G. Di Battista and U. Zwick, eds. *Algorithms – ESA 2003, Lecture Notes in Computaer Science vol. 2832*, Springer-Verlag, Berlin, 2003, pp 196–207.

[15] L. Gao, Y. Zhang, *Computational experience with Lenstra's algorithm*, Technical report TR02-12, Department of Computational and Applied Mathematics, Rice University, Houston, TX, 2002.

[16] O. Goldreich, S. Goldwasser, *On the limits of non-approximability of lattice problems*, Proceedings of the 30th Annual ACM Symposium on Theory of Computing, ACM Press, New York, 1998, pp. 1–9.

[17] M. Grötschel, L. Lovász, and A. Schrijver, *Geometric Algorithms and Combinatorial Optimization*, Springer-Verlag, Berlin, 1988.

[18] C. Hermite, *Extraits de lettres de M. Ch. Hermite à M. Jacobi sur différents objets de la théorie des nombres*, Journal für die reine und angewandte Mathematik, **40**, 1850.

[19] R. Kannan. *Algorithmic geometry of numbers*, Annual Review of Computer Science **2** (1987), 231–267.

[20] R. Kannan, *Minkowski's convex body theorem and integer programming*, Mathematics of Operations Research **12(3)** (1987), 415–440.

[21] R. M. Karp, *Reducibility among combinatorial problems*, Complexity of Computer Computations (R. E. Miller and J. W. Thatcher, (eds.)), Plenum Press, New York, NY, 1972, pp. 85–103.

[22] A. Khinchine, *A quantitative formulation of Kronecker's theory of approximation* (in russian), Izvestiya Akademii Nauk SSR Seriya Matematika **12** (1948), 113–122.

[23] A. K. Lenstra, H. W. Lenstra, Jr., and L. Lovász, *Factoring polynomials with rational coefficients*, Mathematische Annalen **261** (1982), 515–534.

[24] H. W. Lenstra, Jr., *Integer programming with a fixed number of variables*, Report 81-03, Department of Mathematics, University of Amsterdam, 1981.

[25] H. W. Lenstra, Jr., *Integer programming with a fixed number of variables*, Mathematics of Operations Research **8** (1983), 538–548.

[26] LiDIA – A Library for Computational Number Theory. TH Darmstadt/Universität des Saarlandes, Fachbereich Informatik, Institut für Theoretische Informatik. `http://www.informatik.th-darmstadt.de/pub/TI/LiDIA`.

[27] L. Lovász and H. E. Scarf, *The generalized basis reduction algorithm*, Mathematics of Operations Research **17(3)** (1992), 751 – 764.

[28] H. Minkowski, *Über die positiven quadratischen Formen und über kettenbruchänliche Algorithmen*, Journal für die reine und angewandte Mathematik **107** (1891), 278–297.

[29] G. L. Nemhauser and L. A. Wolsey, *Integer and Combinatorial Optimization*, John Wiley & Sons, New York, 1988.

[30] P. Q. Nguyen and J. Stern, *The two faces of lattices in cryptology*, Cryptography and Lattices, International Conference, CaLC 2001, Lecture Notes in Computer Science vol. 2146, Springer-Verlag, Berlin, 2001, pp. 146–180.

[31] C.-P. Schnorr, *A hierarchy of polynomial time lattice basis reduction algorithms*, Theoretical Computer Science **53(2-3)** (1987), 201–224.

[32] C.-P. Schnorr, *Block reduced lattice bases and successive minima*, Combinatorics Probability and Computing **3(4)** (1994), 507–522.

[33] C.-P. Schnorr and M. Euchner, *Lattice basis reduction: improved practical algorithms and solving subset sum problems*, Mathematical Programming **66(2)** (1994), 181–199.

[34] C.-P. Schnorr and H. H. Hörner, *Attacking the Chor-Rivest cryptosystem by improved lattice reduction*, Advances in Cryptology—EUROCRYPT '95, Lecture Notes in Computer Science, vol. 921, Springer-Verlag, Berlin, 1995, pp. 1–12.

[35] A. Schrijver, *Theory of linear and integer programming*, Wiley, Chichester, 1986.

[36] M. Seysen, *Simultaneous reduction of a lattice basis and its reciprocal basis*, Combinatorica **13(3)** (1993), 363–376.

[37] V. Shoup, *NTL: A Library for doing Number Theory*, Courant Institute, New York. `http://www.shoup.net/`.

School of Industrial and Systems Engineering, Georgia Institute of Technology, Atlanta, GA, USA, and Centrum voor Wiskunde en Informatica, Amsterdam, The Netherlands

E-mail address: Karen.Aardal@cwi.nl

Proceedings of Symposia in Applied Mathematics
Volume **61**, 2004

Polyhedral Methods in Discrete Optimization

Alper Atamtürk

ABSTRACT. In the last decade our capability of solving integer programming problems has increased dramatically due to the effectiveness of cutting plane methods based on polyhedral investigations. Polyhedral cutting planes have become central features in optimization software packages for integer programming. Here we present some of the important polyhedral methods used in discrete optimization. We discuss applications to knapsack problems and robust combinatorial optimization.

1. Introduction

A mixed–integer program is a linear optimization problem in which some of the decision variables take discrete values. Mixed–integer programming is a powerful and versatile modeling and optimization framework with diverse applications ranging from radiotherapy to financial/commodity exchanges, from fiberoptic network design to power generation. Rapid advances in information technologies lead to the emergence of new applications of mixed–integer programming with millions of discrete decision variables.

In the last decade we have observed a dramatic improvement in our ability to solve large–scale instances of mixed–integer programs. This achievement has been arguably due to efficient implementations of polyhedral methods in mixed–integer programming software packages. Our objective is to introduce some of these methods. We illustrate applications to knapsack problems and robust combinatorial optimization.

A (linear) mixed–integer program (MIP) is an optimization problem of the form

$$\max\left\{cx + dy : \; Ax + Dy \le b, \; x \in \mathbb{Z}^n_+, \; y \in \mathbb{R}^p_+\right\}, \tag{1.1}$$

where A is an $m \times n$ and D is an $m \times n$ rational matrix and b is an m–dimensional rational column vector. Here $\mathbb{Z}^n_+$ denotes the set of n–dimensional nonnegative integer vectors and $\mathbb{R}^p_+$ denotes the set of p–dimensional nonnegative real vectors. A point $(x, y) \in \mathbb{Z}^n_+ \times \mathbb{R}^p_+$ is said to be feasible if it satisfies the constraints $Ax+Dy \le b$ of MIP. The objective of the problem is to maximize the linear function $cx+dy$ over the set of feasible points. For a thorough introduction to integer programming, the reader is referred to excellent texts [**NW88, Sch87, Wol98**] on this topic.

1991 *Mathematics Subject Classification.* 90C10, 90C11, 90C57.
Key words and phrases. Mixed–integer programming, polyhedra, cutting plane algorithms.
The author was supported in part by NSF Grants DMI–0070127 and DMI–0218265.
March 11, 2004.

2. The polyhedral approach

Let S denote the feasible set of MIP and P denote the linear programming (LP) relaxation of S, i.e.,

$$S := \left\{x \in \mathbb{Z}^n_+,\ y \in \mathbb{R}^p_+ :\ Ax + Dy \leq b\right\},$$

$$P := \left\{x \in \mathbb{R}^n_+,\ y \in \mathbb{R}^p_+ :\ Ax + Dy \leq b\right\}.$$

Let $\text{conv}(S)$ denote the convex hull of S. Since A, D, and b are rational, P is a rational polyhedron, that is, the intersection of a finite number of half–spaces defined by inequalities with rational coefficients. As any polyhedron is convex, P is a convex relaxation of the typically nonconvex feasible set S. Whereas the mixed–integer programming problem $\max\{cx + dy : (x, y) \in S\}$ is $\mathcal{NP}$–hard [**GJ79**], there is a polynomial algorithm for solving the linear programming problem $\max\{cx+dy : (x, y) \in P\}$ [**Kha79**].

The polyhedral approach to mixed–integer programming is an outer convexification scheme, which capitalizes on the fact that convex relaxations—more precisely polyhedral relaxations— of mixed–integer programming are easier to solve than MIP. In this approach one attempts to solve successively stronger linear programming relaxations of MIP. The fundamental result that polyhedral methods rely on is the following theorem.

THEOREM 2.1. [**Mey74**] $\text{conv}(S)$ *is a rational polyhedron.*

The linearity of the objective function of MIP implies that

$$\max\{cx + dy : (x, y) \in S\} = \max\{cx + dy : (x, y) \in \text{conv}(S)\}. \tag{2.1}$$

Equality (2.1) and Theorem 2.1 imply that the nonconvex MIP boils down to solving an equivalent (convex) linear programming problem. However, in general, describing the constraints of $\text{conv}(S)$ as an explicit system of linear inequalities is difficult [**KP80**]. The reader is referred to [**Sch03**] for a thorough coverage of polyhedral results on combinatorial optimization problems.

2.1. Valid inequalities. In this section, we present a general method for deriving linear constraints for $\text{conv}(S)$. Later, in Sections 3 and 4, we discuss how to find constraints for two specific applications.

DEFINITION 2.2. An inequality $\pi x + \sigma y \leq \pi_o$ is said to be *valid* for S if $\pi\bar{x} + \sigma\bar{y} \leq \pi_o$ is satisfied for all $(\bar{x}, \bar{y}) \in S$.

For the *pure integer* case, i.e., when $p = 0$, Chvátal [**Chv73**] gives a finite, but not necessarily efficient, procedure for describing the inequalities of $\text{conv}(S)$, which we summarize here: For any $\mu \in \mathbb{R}^m_+$, the *Chvátal–Gomory* (CG) inequality

$$\sum_{i=1}^{n} \lfloor \mu A_i \rfloor x_i \leq \lfloor \mu b \rfloor, \tag{2.2}$$

where A_i denotes the ith column of A, is a valid for S. Validity of (2.2) follows simply from the integrality of the left–hand side for integral x. Adding to P all CG inequalities, we obtain the so–called *elementary closure* of P:

$$E(P) := \left\{x \in P : \sum_{i=1}^{n} \lfloor \mu A_i \rfloor x_i \leq \lfloor \mu b \rfloor,\ \ \forall \mu \in \mathbb{R}^m_+\right\}.$$

Since CG inequalities are valid for S, we have $\text{conv}(S) \subseteq E(P)$. It turns out that a finite number of μ is sufficient to describe $E(P)$. Consequently, $E(P)$ is a polyhedron itself, which suggests that $E(E(P))$ is well–defined. Chvátal shows that applying the elementary closure iteratively for a finite number of times, one obtains the convex hull of S.

THEOREM 2.3. [**Chv73, Sch87**] *Let P be a rational (or bounded) polyhedron and $P^{(k+1)} = E(P^{(k)})$, where $P^{(1)} = E(P)$. Then $P^{(k)}$ is a rational (or bounded) polyhedron for all $k \geq 1$ and $\text{conv}(S) = P^{(t)}$ for some integer t.*

Chvátal's result is for bounded polyhedra; Schrijver's result is for rational polyhedra. Inequalities (2.2) are called Chvátal–Gomory inequalities because they are equivalent to Gomory's earlier *fractional cuts* [**Gom58, Gom60b**].

For the mixed–integer case, i.e., when $p > 0$, the CG inequality is not valid for S, since due to continuous variables y, the left–hand side of (2.2) is not integral for all points in S. For this general case, CG inequality is extended to the *mixed–integer rounding (MIR) inequality* [**NW90**]

$$\sum_{i=1}^{n}(\lfloor \mu A_i \rfloor + \frac{(f_i - f)^+}{1-f})x_i + \sum_{i=1:\mu G_i<0}^{p} \frac{\mu G_i}{1-f} y_i \leq \lfloor \mu b \rfloor, \tag{2.3}$$

where $f = \mu b - \lfloor \mu b \rfloor$, $f_i = \mu A_i - \lfloor \mu A_i \rfloor$ for $i \in I$, and $\mu \in \mathbb{R}^m_+$, and $(z)^+ = \max\{z, 0\}$. Validity of (2.3) for S, can be shown by a simple disjunctive argument [**MW01**]. Observe that in the special case of pure integer programs ($p = 0$) MIR inequality (2.3) dominates CG inequality (2.2) as $\frac{(f_i - f)^+}{1-f} \geq 0$ for $i = 1, \ldots, n$. MIR inequalities are equivalent to Gomory's earlier *mixed–integer fractional cuts* [**Gom60a**]. Other general classes of inequalities for MIP include *disjunctive inequalities* of Balas [**Bal75a, Bal79**], and *split inequalities* of Cook, Kannan and Schrijver [**CKS90**]. Similar to CG inequalities, MIR and disjunctive inequalities yield the convex hull of S if integer variables are restricted 0–1. Split inequalities are shown to give $\text{conv}(S)$ in a finite number of closure steps when incorporated with a discretization of continuous variables. Sherali and Adams [**SA90**], Lovász and Schrijver [**LS91**], and Lasseree [**Las01**] give other convexification schemes for (mixed) 0–1 programming.

If our goal is to solve the optimization problem (2.1), however, constructing a complete linear description of $\text{conv}(S)$ is neither necessary nor desirable. A cutting plane algorithm for (2.1) solves a sequence of linear programming relaxations, by adding at each iteration one or more new linear constraints that cut off an optimal solution of the previous relaxation, but not the feasible solutions of MIP. Algorithm 1 is a rudimentary cutting plane algorithm.

Gomory's fractional cutting plane algorithm for pure integer programming can be implemented so that it terminates in a finite number of steps. No finite cutting plane algorithm is known for mixed–integer programming. A numerical example of a cutting plane algorithm is given in Section 3.3. A computer program output of a cutting plane algorithm is presented in Section 4.4.

The crucial part of Algorithm 1 is how to find valid inequalities that cut off fractional solutions, which is discussed next.

2.2. Separation. The concept of linear separation is fundamental in polyhedral methods. Let $K \subset \mathbb{R}^n$ be a closed, convex set and $x^* \in \mathbb{R}^n$. Then either $x^* \in K$ or there exits $\pi \in \mathbb{R}^n$ such that $\pi x^* > \pi_o = \max\{\pi x : x \in K\}$ [**Roc70**]. The *separation problem* for a rational polyhedron P is to decide whether a given

Algorithm 1 A cutting plane algorithm.

0. **Initialize.**
   $k \leftarrow 1$ and $P^k \leftarrow P$.
1. **Solve LP relaxation.**
   Let (x^k, y^k) be an optimal solution to the LP $\max\{cx + dy : (x, y) \in P^k\}$.
   If x^k is integer, stop; (x^k, y^k) is an optimal solution to MIP.
2. **Add cut.** (x^k has a fractional component)
   Find a valid inequality $\pi x + \sigma y \leq \pi_o$ for S that cuts off (x^k, y^k).
   $P^{k+1} \leftarrow P^k \cap \{(x, y) : \pi x + \sigma y \leq \pi_o\}$ and $k \leftarrow k + 1$.
   Go to Step 1.

rational point x^* is in P or not, and in the latter case, to construct a valid inequality $\pi x \leq \pi_o$ for P such that $\pi_o < \pi x^*$. There is a polynomial algorithm for solving the separation problem for P if and only if there is a polynomial algorithm for solving the optimization problem $\max\{cx : x \in P\}$ for rational $c \in \mathbb{R}^n$ [**GLS93**].

Since the optimization problem (2.1) is $\mathcal{NP}$–hard, in general, there is no polynomial algorithm for finding a valid inequality for conv(S) that cuts off a given point (x, y), unless $\mathcal{P} = \mathcal{NP}$. However, finding a cut may be easy for specific points (x, y), as in Gomory's fractional cutting plane algorithm [**Gom60b**]. In practice, polynomial heuristics are employed to find cuts. A prevalent approach is to use a relaxation $T \supseteq \text{conv}(S)$ that is easier to separate. If a cut separates T and (x, y), then it separates conv(S) and (x, y). Since the converse statement does not hold, this approach may fail to find a cut. In that case, one either tries a different relaxation or resorts to enumeration.

State–of–the–art optimization software systems such as CPLEX[1], LINDO[2], and XPRESS–MP[3] integrate cutting plane algorithm and enumeration to solve MIPs. See [**AS03**] for a review that discusses many functionalities of MIP software systems.

2.3. Strong valid inequalities.

DEFINITION 2.4. Given a valid inequality $\pi x + \sigma y \leq \pi_o$ for S, $F := \{(x, y) \in \text{conv}(S) : \pi x + \sigma y = \pi_o\}$ is called the *face* of conv(S) defined by $\pi x + \sigma y \leq \pi_o$. A face F of conv(S) is *proper* if $F \neq \emptyset$ and $F \neq \text{conv}(S)$. Maximal proper faces are called *facets.*

Dimension of the face a valid inequality defines, may be viewed as a measure of the strength of the inequality. Facets of a polyhedron are its highest dimensional proper faces. It is known that facet–defining inequalities of conv(S) are sufficient to describe conv(S) as a system of linear inequalities. Therefore, given a discrete optimization problem, it is desirable to identify inequalities that define high–dimensional faces, or preferably facets, of the convex hull of the feasible set. Even if only a subset of facet–defining inequalities is known, they can be useful as cutting planes in strengthening the LP relaxations when incorporated in LP–based enumeration methods.

[1]CPLEX is a trademark of ILOG, Inc.

[2]LINDO is a trademark of LINDO Systems, Inc.

[3]XPRESS–MP is a trademark of Dash Optimization.

An effective method for deriving strong valid inequalities is *lifting*. Lifting refers to extending simple valid inequalities for low–dimensional restrictions of polyhedra to ones that are valid in high dimensions. Lifting of inequalities has been introduced by Gomory [**Gom69**] in the context of the group problem. Padberg [**Pad79**] described the sequential lifting procedure for 0–1 programming, i.e., when integer variables are allowed to take one of the two values: zero and one. Since then, it has been studied extensively [**Ata01, BZ78, GNS98, GNS00, JP81, LW03, Pad73, RdN02, Wol76, Wol77, Zem78, Zem89**], particularly, for 0–1 and mixed 0–1 programming problems. Below we summarize the use of superadditive functions for lifting inequalities for general MIPs. Details can be found in [**Ata01**].

We define $[i,k] := \{j \in \mathbb{Z} : i \leq j \leq k\}$. For a subset $T \subseteq [1,n]$, let A_T denote the matrix of columns A_i, $i \in T$ of A. Let ℓ_i (u_i) be the smallest (largest) value x_i, $i \in [1,n]$ takes in any feasible solution to S. Let (L, U, R) be a partition of $[1,n]$ with $r = |R|$ so that $u_i < \infty$ for all $i \in U$ and

$$S(L,U) := \left\{ x_R \in \mathbb{Z}^r_+, y \in \mathbb{R}^p_+ : A_R x_R + Dy \leq d := b - A_L \ell_L - A_U u_U \right\},$$

i.e., the restriction of S obtained by fixing $x_i = \ell_i$, $i \in L$ and $x_i = u_i$, $i \in U$, is nonempty. For a valid inequality

$$\pi_R x_R + \sigma y \leq \pi_o \tag{2.4}$$

for $S(L,U)$, define the *lifting function* $\Phi : \mathbb{R}^m \mapsto \mathbb{R} \cup \{\infty\}$ as

$$\Phi(a) = \pi_o - \max\left\{ \pi_R x_R + \sigma y : A_R x_R + Dy \leq d - a,\ x_R \in \mathbb{Z}^r_+,\ y \in \mathbb{R}^p_+ \right\}.$$

We let $\Phi(a) = \infty$ if the optimization problem above, referred as the *lifting problem*, is infeasible. Since (2.4) is valid for $S(L,U)$, the lifting problem is bounded for all $a \in \mathbb{R}^m$ for which it is feasible. The lifting function Φ is useful in proving the validity of extensions of (2.4) to higher dimensions. Observe that

$$\pi_R x_R + \pi_L(x_L - \ell_L) + \pi_U(u_U - x_U) + \sigma y \leq \pi_o \tag{2.5}$$

is valid for S if and only if

$$\pi_L(x_L - \ell_L) + \pi_U(u_U - x_U) \leq \Phi(A_L(x_L - \ell_L) + A_U(x_U - u_U)) \tag{2.6}$$

for all $(x,y) \in S$. Now, suppose that $\phi : \mathbb{R}^m \mapsto \mathbb{R}$ is a *superadditive* lower bound on Φ, that is, $\phi(a_1) \leq \Phi(a_1)$ and $\phi(a_1) + \phi(a_2) \leq \phi(a_1 + a_2)$ for all $a_1, a_2 \in \mathbb{R}^m$. Then, since

$$\begin{aligned}\sum_{i\in L} \phi(A_i)(x_i - \ell_i) - \sum_{i \in U} \phi(-A_i)(x_i - u_i) &\leq \phi(A_L(x_L - \ell_L) + A_U(x_U - u_U)) \\ &\leq \Phi(A_L(x_L - \ell_L) + A_U(x_U - u_U)),\end{aligned}$$

inequality (2.5) is valid for S with $\pi_i = \phi(A_i)$, $i \in L$ and $\pi_i = \phi(-A_i)$, $i \in U$. In other words, using a superadditive lower bound ϕ on Φ, inequality (2.4) can be extended to a valid inequality

$$\pi_R x_R + \sum_{i \in L} \phi(A_i)(x_i - \ell_i) + \sum_{i \in U} \phi(-A_i)(u_i - x_i) + \sigma y \leq \pi_o \tag{2.7}$$

for S. Furthermore, if $\phi(A_i) = \Phi(A_i)$ for all $i \in L$, $\phi(-A_i) = \Phi(-A_i)$ for all $i \in U$ and inequality (2.4) defines a k–dimensional face of $\text{conv}(S(L,U))$, then inequality (2.7) defines a face of $\text{conv}(S)$ of dimension at least $k + |L| + |U|$ [**Ata01**].

2.4. Disjunctive formulations. The disjunctive approach is another useful way for identifying valid inequalities and strong LP relaxations for mixed–integer programming. Let $P_k = \{x \in \mathbb{R}^n_+ : A^k x \leq b^k\}$ $(\neq \emptyset)$ for $k \in [1, q]$. Suppose we are interested in optimizing a linear function over the disjunction (union) of these polyhedra

$$P := \cup_{k=1}^q P_k,$$

which is typically a nonconvex set. From the linearity of the objective function it follows that the optimization problem can equivalently be solved over conv(P). This statement is a generalization of the equality (2.1) since by fixing all integer variables to feasible values we obtain polyhedral restrictions of S. The following result is a characterization of conv(P) using auxiliary variables.

THEOREM 2.5. [**Bal79**] *Let P_k, $k \in [1, q]$ and P be defined as above. Then*

$$\operatorname{conv}(P) = \operatorname{proj}_x(Q), \text{ where } Q := \left\{ \begin{array}{l} x \in \mathbb{R}^n_+, w \in \mathbb{R}^{qn}_+, \lambda \in \mathbb{R}^q_+ : \\ \mathbf{1}'\lambda = 1 \\ A^k w_k \leq b^k \lambda_k, \quad k \in [1, q] \\ x = \sum_{k=1}^q w_k \end{array} \right\}.$$

In particular, λ is integral in extreme points of Q.

We will illustrate the disjunctive approach for deriving strong LP relaxations in Section 4. Below we give a simple example.

EXAMPLE 2.6. Suppose the feasible set of an optimization problem is given as

$$P = \{ \ x \in \mathbb{R}^n_+ : \text{either } a^1 x \leq 1 \text{ or } a^2 x \leq 1 \ \}$$

with $a^1, a^2 \geq 0$. P is, in general, a nonconvex set. This type of disjunctive constraint frequently arises in scheduling problems and is typically formulated as a mixed 0–1 program by introducing an auxiliary binary variable z:

$$S = \left\{x \in \mathbb{R}^n_+, z \in \{0, 1\} : a^1 x \leq 1 + M(1 - z), \ a^2 x \leq 1 + Mz\right\},$$

where M is a constant such that $M \geq \max\{a^1 x - 1, a^2 x - 1 : x \in \mathbb{R}^n_+\}$. Constraint $a^1 x \leq 1$ holds if $z = 0$, whereas constraint $a^2 x \leq 1$ holds if $z = 1$. The LP relaxation of S is generally weak. However, using Theorem 2.5, we can define the convex hull of P implicitly as

$$\operatorname{conv}(P) = \operatorname{proj}_x \left\{ \begin{array}{l} x \in \mathbb{R}^n_+, w \in \mathbb{R}^{2n}_+, \lambda \in \mathbb{R}^2_+ : \\ \lambda_1 + \lambda_2 = 1 \\ a^1 w_1 \leq \lambda_1, \ a^1 w_2 \leq \lambda_2 \\ x = w_1 + w_2 \end{array} \right\}.$$

Projecting out the auxiliary variables w and λ, it can be shown that

$$\operatorname{conv}(P) = \{ \ x \in \mathbb{R}^n_+ : \sum_{i=1}^n \min\{a^1_i, a^2_i\} x_i \leq 1 \ \}.$$

In general, writing an explicit description of the convex hull of P is difficult.

3. Application to the knapsack problem

The knapsack problem is an integer program defined over a single "budget" constraint. Formally it is stated as

$$\max \; \{ \; cx : ax \leq b, \; 0 \leq x \leq u, \; x \in \mathbb{Z}^n \; \} .$$

If necessary, by complementing variables, i.e., replacing x_i with $u_i - \bar{x}_i$, we may assume without loss of generality that $a \geq 0$. We may also assume that $0 < a_i \leq b$ for all i, since if $a_i = 0$, the value of x_i in an optimal solution can be determined from the objective coefficient c_i, or if $a_i > b_i$, then $x_i = 0$ in every feasible solution.

The knapsack problem is $\mathcal{NP}$–hard [**GJ79**] and it frequently arises as a subproblem of many practical problems. Since every constraint of an integer program without continuous variables can be viewed as a budget constraint, the feasible set of the knapsack problem forms a relaxation for any integer program. Consequently, valid inequalities for the knapsack problem are valid for any integer program and they can be effective in strengthening LP relaxations of general integer programs. For this reason strong valid inequalities for the knapsack problem play an important role in general purpose MIP software packages.

3.1. Binary knapsacks. In this section we describe some of the important classes of facets of the binary knapsack polytope

$$K_B := \text{conv} \{ \; x \in \{0,1\}^n : \; ax \leq b \; \} .$$

A subset C of the index set of variables $N := [1, n]$ is called a *cover* if $\lambda = a(C) - b > 0$, where $a(C) = \sum_{i \in C} a_i$.

For a cover C, let us consider the restriction $K_B(N \setminus C)$ of K_B, obtained by fixing all x_i, $i \in N \setminus C$ to zero. Since the sum of the coefficients a_i, $i \in C$ exceeds the knapsack capacity by $\lambda > 0$, all variables x_i, $i \in C$ cannot be one simultaneously in a feasible solution to $K_B(N \setminus C)$. Therefore, the *cover inequality* [**Bal75b, HJP75, Wol75**]

$$\sum_{i \in C} x_i \leq |C| - 1 \tag{3.1}$$

is valid for $K_B(N \setminus C)$. Cover inequality (3.1) defines a facet of $K_B(N \setminus C)$ if and only if C is a minimal cover, that is, $a(C \setminus \{i\}) \leq b$ for all $i \in C$.

Lifting of cover inequalities has been studied [**Bal75b, BZ78, BZ84, GNS98, GNS99, HJP75, Wol75, Zem89**] for extending them to facet–defining inequalities for K_B. Here we illustrate superadditive lifting of cover inequalities [**GNS00**]. Suppose the cover inequality (3.1) defines a facet of $K_B(N \setminus C)$; thus, $a_i \geq \lambda$ for all $i \in C$. For convenience we write the inequality as

$$\sum_{i \in C} \lambda x_i \leq \lambda(|C| - 1). \tag{3.2}$$

Suppose (w.l.o.g.) that $C = \{1, 2, \ldots, |C|\}$ and $a_1 \geq a_2 \geq \cdots \geq a_{|C|}$. Let $A_i = \sum_{k=1}^{i} a_k$ for $i \in \{1, 2, \ldots, |C|\}$ and $A_0 = 0$. Then the lifting function of (3.2)

$$\Theta(a) = \lambda(|C| - 1) - \max \left\{ \sum_{i \in C} \lambda x_i : \sum_{i \in C} a_i x_i \leq b - a \right\}$$

for $a \geq 0$ can be expressed in closed form as

$$\Theta(a) = \begin{cases} 0 & \text{if } A_0 \leq a \leq A_1 - \lambda, \\ i\lambda & \text{if } A_i - \lambda < a \leq A_{i+1} - \lambda, \\ \infty & \text{if } a > b, \end{cases}$$

where $i \in \{0, 1, \ldots, |C|-1\}$. Θ is a step function and is generally not superadditive. However, letting $\{i \in C : a_i > \lambda\} = \{1, 2, \ldots, \ell\}$, it is easy to check that the continuous function $\varphi : \mathbb{R}_+ \mapsto \mathbb{R}_+$ defined as

$$\varphi(a) = \begin{cases} i\lambda & \text{if } A_i \leq a \leq A_{i+1} - \lambda, \\ i\lambda + (a - A_i) & \text{if } A_i - \lambda \leq a \leq A_i, \\ \ell\lambda + (a - A_\ell) & \text{if } A_\ell - \lambda \leq a, \end{cases}$$

where $i \in \{0, 1, \ldots, \ell-1\}$, is a superadditive lower bound on Θ for $a \geq 0$. Therefore,

$$\sum_{i \in C} x_i + \sum_{i \in N \setminus C} \frac{\varphi(a_i)}{\lambda} x_i \leq |C| - 1 \tag{3.3}$$

is a valid inequality for K_B. Moreover, since $\varphi(a) = \Theta(a)$ for $A_i \leq a \leq A_{i+1} - \lambda$, inequality (3.3) is facet–defining for K_B if for all $k \in N \setminus C$ we have $A_j \leq a_k \leq A_{j+1} - \lambda$ for some $j \in \{0, 1, \ldots, \ell - 1\}$. This result implies the partial characterization of the exact lifting coefficients by Balas and Zemel [**BZ78**]. Gu et al. [**GNS00**] describe a stronger superadditive lower bound than φ for Θ. Marchand and Wolsey [**MW99**] generalize inequalities (3.3) for the mixed 0–1 knapsack problem. See [**Ata03a**] for other strong inequalities for the knapsack polytope.

3.2. Integer knapsacks. In this section we describe valid inequalities for the the integer knapsack polytope

$$K_I := conv\{\ x \in \mathbb{Z}^n_+ : ax \leq b, \quad x \leq u\ \}.$$

Let us generalize the definition of a cover for integer knapsacks. We say that $C \subseteq N = [1, n]$ is a *cover* if $\lambda = \sum_{i \in C} u_i a_i - b > 0$; therefore, C is a minimal cover if $\sum_{i \in C \setminus \{k\}} u_i a_i \leq b$ for all $k \in C$.

Now consider the restriction $K_I(N \setminus C)$ of K_I obtained by fixing all x_i, $i \in N \setminus C$ to zero. Extending the combinatorial argument on covers for the binary knapsack set, since not all variables x_i, $i \in C$ can be at their upper bound simultaneously, inequality

$$\sum_{i \in C} x_i \leq u(C) - 1 \tag{3.4}$$

is valid for $K_I(N \setminus C)$. However, unlike (3.1), inequality (3.4) is not facet–defining and may not even define a proper face of $K_I(N \setminus C)$ even if C is a minimal cover.

Another way to write inequality (3.4) is

$$\sum_{i \in C} (u_i - x_i) \geq 1, \tag{3.5}$$

that is, at least one x_i must be less than its upper bound. However, observe that if $a_i < \lambda$ for all $i \in C$, the left–hand side of (3.5) is greater than one for all feasible solutions, which suggests the strengthening

$$\sum_{i \in C} (u_i - x_i) \geq \lceil \lambda / \bar{a} \rceil, \tag{3.6}$$

where $\bar{a} = \max_{i \in C} a_i$ (see [**CCMW98**]). Inequality (3.6) is indeed a CG inequality. To see this, write $\sum_{i \in C} a_i x_i \leq b$ as $\sum_{i \in C} -a_i(u_i - x_i) \leq -\lambda$. Multiplying the latter inequality by $1/\bar{a} > 0$ and rounding gives (3.6). For the 0–1 case, i.e., when $u = \mathbf{1}$, if C is a minimal cover, inequality (3.6) equals the cover inequality (3.1). Consider the example below.

EXAMPLE 3.1. Suppose the integer knapsack polytope is given as

$$K_I = conv\{x \in \mathbb{Z}_+^2 : 5x_1 + 9x_2 \leq 45, \ x_1 \leq 6, \ x_2 \leq 4\}.$$

Let $C = \{1, 2\}$ be a cover with $\lambda = 30 + 36 - 45 = 21$. As shown in Figure 1, inequality (3.5)

$$(6 - x_1) + (4 - x_2) \geq 1 \quad \text{or} \quad x_1 + x_2 \leq 9$$

does not support K_I. However, the strengthened inequality (3.6)

$$(6 - x_1) + (4 - x_2) \geq \lceil 21/9 \rceil = 3 \quad \text{or} \quad x_1 + x_2 \leq 7$$

defines a facet of K_I in this example.

For binary knapsacks, whenever C is a minimal cover, the cover inequality (3.1) defines a facet of $K_B(N \setminus C)$. Furthermore, it is also easy to see that, if C is a minimal cover, the bounds on the variables and the cover inequality give a complete description of $K_B(N \setminus C)$.

As seen in Figure 1, $K_I(N \setminus C)$ has other non–trivial facets besides the one defined by $x_1 + x_2 \leq 7$. Also it should be clear that, in general, $K_I(N \setminus C)$ may not have any facet of the form $\sum_{i \in C} x_i \leq \pi_o$. Therefore, we describe a generalization of inequalities (3.6), having coefficients that are not restricted to zero and one. Let $C \subseteq N$ be a cover. For any $\rho > 0$, the *integer cover inequality* [**Ata03a**]

$$\sum_{i \in C} \lceil \min\{a_i, \lambda\}/\rho \rceil (u_i - x_i) \geq \lceil \lambda/\rho \rceil \tag{3.7}$$

is valid for K_I.

It is of interest to know when inequalities (3.7) define facets of $K_I(N \setminus C)$. A necessary and sufficient condition for a subclass of the integer cover inequalities to define facets of $K_I(N \setminus C)$ is stated in the next theorem. For a cover C and $i, \ell \in C$, let $\kappa_{i\ell} = \lceil \min\{a_i, \lambda\}/a_\ell \rceil$ and consider the integer cover inequality

$$\sum_{i \in C} \kappa_{i\ell}(u_i - x_i) \geq \lceil \lambda/a_\ell \rceil. \tag{3.8}$$

THEOREM 3.2. [**Ata03a**] *Let $C \subseteq N$ be a cover and $\ell \in C$ be such that $\mu = u_\ell a_\ell - \lambda \geq 0$. The integer cover inequality (3.8) is facet-defining for $K_I(N \setminus C)$ if and only if $a_i \geq \min\{\lambda, \kappa_{i\ell} a_\ell - r\}$ for all $i \in C \setminus \{\ell\}$, where $r = \mu - \lfloor \mu/a_\ell \rfloor a_\ell$.*

The condition of Theorem 3.2 generalizes the facet condition of minimality of covers for binary knapsacks to integer knapsacks. To see this, observe that for the binary knapsack set, we have $\mu = a_\ell - \lambda < a_\ell$ and $r = \mu$. Since $\mu \geq 0$ implies $k_{i\ell} = 1$ for all $i \in C$, Theorem 3.2 reduces to stating that $\sum_{i \in C} x_i \leq |C| - 1$ is facet–defining for $K_B(N \setminus C)$ if and only if $a_i \geq \lambda$ for all $i \in C$, that is, C is a minimal cover.

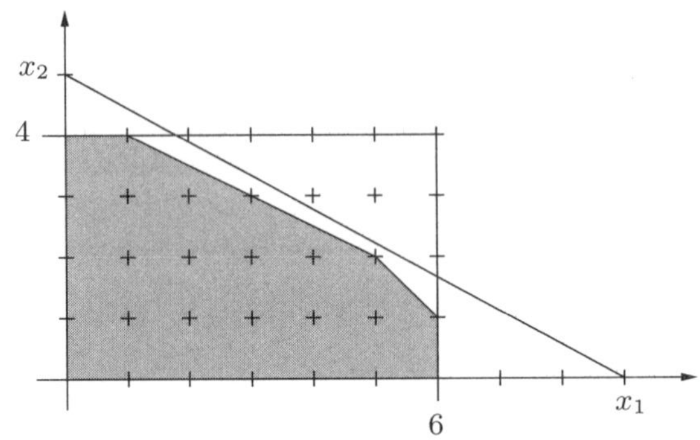

FIGURE 1. Facets of the integer knapsack set

EXAMPLE 3.3. Let $C = \{1, 2\}$ be a cover for K_I defined in Example 3.1 with $\lambda = 30 + 36 - 45 = 21$. For $\ell = 1$, we have $\mu = 30 - 21 = 9$ and $r = 4$. Then since $\lceil \lambda/a_1 \rceil = \lceil 21/5 \rceil = 5$, $\kappa_{11} = 1$ and $\kappa_{21} = \lceil 9/5 \rceil = 2$, the corresponding integer cover inequality is

$$(6 - x_1) + 2(4 - x_2) \geq 5 \quad \text{or} \quad x_1 + 2x_2 \leq 9.$$

On the other hand, for $\ell = 2$, we have $\mu = 36 - 21 = 15$ and $r = 6$. In this case, since $\lceil \lambda/a_2 \rceil = \lceil 21/9 \rceil = 3$, $\kappa_{12} = \lceil 5/9 \rceil = 1$ and $\kappa_{22} = 1$, the integer cover inequality is

$$(6 - x_1) + (4 - x_2) \geq 3 \quad \text{or} \quad x_1 + x_2 \leq 7.$$

Both of these inequalities satisfy the condition of Theorem 3.2, and hence define facets of K_I. This is also illustrated in Figure 1.

Integer cover inequalities (3.8) can be strengthened by introducing nonpositive coefficients for variables not in the cover using the lifting functions defined in [**Ata03b**].

3.3. A numerical cutting plane example based on cover inequalities. Here we demonstrate a cutting plane algorithm based on cover inequalities (3.1) for the binary knapsack problem. Consider the example with four variables:

$$\begin{aligned} z_{IP} = \max\ & 16x_1 + 22x_2 + 12x_3 + 8x_4 \\ \text{s.t.}\ & 5x_1 + 7x_2 + 4x_3 + 3x_4 \leq 14 \\ & x_i \in \{0, 1\},\ i = 1, 2, 3, 4. \end{aligned} \tag{3.9}$$

Step 1. First we solve the LP relaxation of (3.9) by relaxing the domain of the variables as $0 \leq x_i \leq 1$, $i = 1, 2, 3, 4$. An optimal solution for the LP relaxation is $x^1_{LP} = (1, 1, \frac{1}{2}, 0)$ with objective value $z^1_{LP} = 44$.

Step 2. Since the cover inequality

$$x_1 + x_2 + x_3 \leq 2$$

corresponding to minimal cover $C = \{1, 2, 3\}$ cuts off x^1_{LP}, we add it to the LP relaxation and resolve. The new solution turns out to be $x^2_{LP} = (1, 1, 0, \frac{2}{3})$ with an improved objective value $z^2_{LP} = 43\frac{1}{3}$.

Step 3. This time the cover inequality

$$x_1 + x_2 + x_4 \leq 2$$

corresponding to minimal cover $C = \{1, 2, 4\}$ cuts off x^2_{LP}. Therefore, we add it to the LP relaxation and resolve. The optimal solution to this LP relaxation turns out to be integral: $x^3_{LP} = (0, 1, 1, 1)$ with objective value $z^3_{LP} = 42$. Since x^3_{LP} is a feasible (integer) solution to (3.9), which is optimal for an LP relaxation, it is optimal for (3.9) as well.

Observe that as new cuts are added, the LP relaxation is strengthened and its optimal objective value decreases. We should point out that one may not be as lucky as on this example to find an optimal solution just by adding cover inequalities. Actually, finding cover inequalities violated by LP relaxations is itself another optimization problem, which is $\mathcal{NP}$–hard [**KNT98**]; however, there are heuristic algorithms that perform very well in practice [**GNS98**]. If no more cover violated cover inequities can be found, one can resort to adding other classes of known valid inequalities or to branching.

4. Application to robust combinatorial optimization

In this section we illustrate how polyhedral methods may be employed for deriving strong LP relaxations for robust combinatorial optimization problems. Robust optimization is a paradigm for finding a solution to an optimization problem when the parameters of the problem are not fixed, but belong to a well–defined uncertainty set. In this scheme, one typically aims for a solution that maximizes (or minimizes) the worst objective value for all realizations of the parameters in the uncertainty set.

Consider a combinatorial optimization problem

$$\max\{\bar{c}x : x \in \mathcal{F}\} \tag{4.1}$$

where $\mathcal{F} \subseteq \{0,1\}^n$ is the feasible region. Now suppose that objective coefficient of variable x_i is not fixed as $\bar{c}_i$, but lies in the uncertainty interval $c_i - d_i \leq \bar{c}_i \leq c_i$, where $d_i \geq 0$. Without loss of generality, we assume that the variables are indexed so that $d_0 := 0 \leq d_1 \leq \ldots \leq d_n$. Bertsimas and Sim [**BS03**] define the robust counterpart of (4.1) as

$$\max\left\{cx - \max_{S\subseteq[1,n],|S|\leq r} \sum_{i\in S} d_i x_i : \ x \in \mathcal{F}\right\}. \tag{4.2}$$

Here r is a parameter used to control the degree of conservatism. The bigger the r is, the more conservative is the solution. One obtains the nominal problem (4.1) as a special case by setting $r = 0$. After linearizing the objective using the LP dual of the inner maximization problem, (4.2) can be rewritten as the following mixed 0–1 program [**BS03**]

$$\max\{cx - \mathbf{1}'y - rz : \ (x, y, z) \in \mathcal{F} \cap \mathcal{R}\} \tag{4.3}$$

where

$$\mathcal{R} := \left\{ \ (x, y, z) \in \{0,1\}^n \times \mathbb{R}^{n+1}_+ : \ y_i + z \geq d_i x_i, \ \ i \in [1, n] \ \right\}.$$

Bertsimas and Sim show that if the nominal problem (4.1) is solvable in polynomial time, so is its robust counterpart (4.3). However, in general, the given robust model (4.3) has typically a very weak LP bound, which makes it difficult to solve with an LP–based MIP solver. Here we describe reformulations of (4.2) with strong LP relaxations.

4.1. A disjunctive formulation. The first strong formulation is based on the observation that the variable z takes at most $n+1$ distinct values in extreme points of conv($\mathcal{R}$). Let $\mathcal{R}(\delta) = \{(x,y,z) \in \mathcal{R} : z = \delta\}$ for a fixed $\delta \geq 0$, that is,

$$\mathcal{R}(\delta) = \left\{ \ (x,y,z) \in \{0,1\}^n \times \mathbb{R}^n_+ \times \delta : d_i x_i \leq y_i + \delta, \ i \in [1,n] \ \right\}.$$

Since the constraints of $\mathcal{R}(\delta)$ are decoupled, by improving the coefficients of each constraint, it is easy to see that the convex hull of $\mathcal{R}(\delta)$ can be written as

$$\text{conv}(\mathcal{R}(\delta)) = \left\{ \ (x,y,z) \in \mathbb{R}^{2n} \times \delta : (d_i - \delta)^+ x_i \leq y_i, \ i \in [1,n], \ \mathbf{0} \leq x \leq \mathbf{1} \ \right\}.$$

On the other hand, since $z \in \{d_0, d_1, \dots, d_n\}$ in extreme points of conv($\mathcal{R}$),

$$\text{conv}(\mathcal{R}) = \mathit{conv}\left(\cup_{k=0}^n \ \text{conv}(\mathcal{R}(d_k)\right) + \mathcal{K},$$

where

$$\mathcal{K} = \{(x,y,z) \in \mathbb{R}^{2n+1} : x = \mathbf{0}, \ y \geq \mathbf{0}, \ z \geq 0\}$$

is the recession cone of the LP relaxation of $\mathcal{R}$. Then, from Theorem 2.5, for

$$\mathcal{D} = \left\{ \begin{array}{ll} (x,y,z,w,\lambda) \in \mathbb{R}^{n^2+4n+2} : & \\ \mathbf{1}'\lambda = 1 & \\ 0 \leq w_{ik} \leq \lambda_k & i \in [1,n] \ k \in [0,n] \\ z \geq \sum_{k=0}^n d_k \lambda_k & \\ y_i \geq \sum_{k=0}^n (d_i - d_k)^+ w_{ik} & i \in [1,n] \\ x = \sum_{k=0}^n w_k & \end{array} \right\}$$

we have conv($\mathcal{R}$) $=$ proj$_{x,y,z}(\mathcal{D})$. Due to the implicit convexification of $\mathcal{R}$ by $\mathcal{D}$, the formulation

$$\max\{cx - \mathbf{1}'y - rz : \ x \in \mathcal{F}, \ (x,y,z,w,\lambda) \in \mathcal{D}\} \tag{4.4}$$

for robust 0–1 programming has a stronger LP relaxation than (4.3).

4.2. Inequalities in the original space. Although polynomial in size, the disjunctive formulation (4.4) has a quadratic number of additional variables and constraints used to model robustness. Therefore, it may be preferable to give an explicit description of conv($\mathcal{R}$) in the original space of variables x, y, z.

THEOREM 4.1. [**Ata03c**] *For $T = \{i_1, i_2, \dots, i_p\} \subseteq [1,n]$ with $0 = d_{i_0} \leq d_{i_1} \leq \dots \leq d_{i_p}$, inequality*

$$\sum_{j=1}^{p} (d_{i_j} - d_{i_{j-1}}) x_{i_j} \leq z + \sum_{i \in T} y_i \tag{4.5}$$

is valid for $\mathcal{R}$. Furthermore, (4.5) *defines a facet of* conv($\mathcal{R}$) *if and only if $0 < d_{i_1} < \dots < d_{i_p}$.*

Separation. Even though conv($\mathcal{R}$) can have as many as 2^n facets defined by inequalities (4.5), there is a polynomial separation algorithm for conv($\mathcal{R}$). Let G be a directed graph with $n+2$ vertices labeled from 0 to $n+1$; and let (i,j) be an arc in G if and only if $0 \leq i < j \leq n+1$. There is a one–to–one correspondence between inequalities (4.5) and the 0–$(n+1)$ paths in G; that is, $j \in T$ if and only if j is contained on a 0–$(n+1)$ path. Given a point $(x,y,z) \in \mathbb{R}^{2n+1}_+$, let the length of arc (i,j) be $y_j - (d_j - d_i)$ if $j \in [1,n]$ and z if $j = n+1$, and ζ be the length of a shortest 0–$(n+1)$ path. Then there exists an inequality (4.5) violated by (x,y,z) if and only if $\zeta < 0$, which can be checked in $O(n^2)$ by finding a shortest path on this acyclic network.

4.3. An extended formulation. Consider the polyhedron

$$\mathcal{P} = \{(x, y, z) \in \mathbb{R}^{2n+1} : \mathbf{0} \leq x \leq \mathbf{1},\ y \geq \mathbf{0} \text{ and } (4.5)\}.$$

Since (4.5) is valid for $\mathcal{R}$, we have $\text{conv}(\mathcal{R}) \subseteq \mathcal{P}$. Is $\mathcal{P} \subseteq \text{conv}(\mathcal{R})$? In order to answer this question, we will define an extended formulation of $\mathcal{P}$ with auxiliary variables. Since the separation problem of inequalities (4.5) is a shortest path problem, it can be formulated as the linear program

$$\begin{aligned}
\min \quad & \sum_{1 \leq i < j \leq n} (y_j - (d_j - d_i)x_j) f_{ij} + \sum_{i=1}^{n} z f_{in+1} \\
\text{s.t.} \quad & \sum_{0 \leq i < j} f_{ij} - \sum_{n+1 \geq i > j} f_{ji} = \begin{cases} 0 & \text{if } j \in [1, n] \\ -1 & \text{if } j = 0 \\ +1 & \text{if } j = n+1 \end{cases} \\
& f \geq \mathbf{0}.
\end{aligned} \tag{4.6}$$

Introducing dual variables v for the equality constraints of (4.6), we can write the constraints of the dual problem as

$$\begin{aligned}
(d_j - d_i)x_j + v_j - v_i \leq y_j & \quad 0 \leq i < j \leq n \\
v_{n+1} - v_i \leq z & \quad 0 \leq i \leq n.
\end{aligned}$$

Furthermore, the objective of the dual $v_{n+1} - v_0$ is nonnegative if and only if no inequality (4.5) is violated by $(x, y, z) \in \mathbb{R}^{2n+1}$. Hence, by construction, $\mathcal{P} = \text{proj}_{x,y,z}(\mathcal{Q})$, where

$$\mathcal{Q} := \left\{ (x, y, z, v) \in \mathbb{R}^{3n+3} : \begin{array}{rl} (d_j - d_i)x_j + v_j - v_i \leq y_j & 0 \leq i < j \leq n \\ v_{n+1} - v_i \leq z & 0 \leq i \leq n \\ v_{n+1} - v_0 \geq 0 & \\ y \geq \mathbf{0} & \\ \mathbf{0} \leq x \leq \mathbf{1} & \end{array} \right\}.$$

It can be shown that every minimal face of $\mathcal{Q}$ contains a point (x, y, z) with integral x, which implies the following theorem.

THEOREM 4.2. [**Ata03c**] $\mathcal{P} = \text{proj}_{x,y,z}(\mathcal{Q}) = \text{proj}_{x,y,z}(\mathcal{D}) = \text{conv}(\mathcal{R})$.

4.4. Solving the robust knapsack problem with CPLEX. In this section we present an implementation of a cutting plane algorithm to solve the robust knapsack problem

$$\max \{cx - \mathbf{1}'y - rz : \ ax \leq b,\ (x, y, z) \in \mathcal{R}\}.$$

Inequalities (4.5) are added as cuts to the above formulation to improve the LP relaxation. The implementation is done using the callable libraries of CPLEX. The code used to input the problem to CPLEX can be found in the Appendix section.

Following is the output of a computer program that solves a robust knapsack problem with 100 binary variables. The `Objective` column shows the improvement of the LP objective as cuts are added to the formulation. The `Cuts` column shows the number of cutting planes added at each iteration.

```
% robust-knapsack
Root relaxation solution time =     0.00 sec.

        Nodes                                            Cuts/
  Node   Left    Objective  IInf  Best Integer     Best Node   ItCnt     Gap
     0     0   7635.1703    86                     7635.1703     118
*   0+     0                 0      1822.0000      7635.1703     118  319.05%
               6890.0616    85      1822.0000    Cuts:  257     147  278.16%
*   0+     0                 0      1837.0000      6890.0616     147  275.07%
               6272.0840    83      1837.0000    Cuts:  170     159  241.43%
               5551.4207    85      1837.0000    Cuts:  166     223  202.20%
               5267.6967    84      1837.0000    Cuts:  116     257  186.76%
               5198.3805    84      1837.0000      User:  83     276  182.98%
               5113.1258    85      1837.0000      User:  83     288  178.34%
*   0+     0                 0      1953.0000      5113.1258     288  161.81%
               4898.2064    83      1953.0000      User:  85     310  150.80%
               4700.0203    86      1953.0000      User:  81     337  140.66%
*   0+     0                 0      2232.0000      4700.0203     337  110.57%
               4501.3912    83      2232.0000      User:  84     377  101.68%
               4317.1583    83      2232.0000      User:  81     406   93.42%
               4181.6975    82      2232.0000      User:  81     432   87.35%
               4071.2503    83      2232.0000      User:  80     487   82.40%
               3887.0125    85      2232.0000      User:  81     547   74.15%
               3604.8825    84      2232.0000      User:  81     629   61.51%
*   0+     0                 0      2311.0000      3604.8825     629   55.99%
               3447.3163    86      2311.0000      User:  84     691   49.17%
               3205.9646    83      2311.0000      User:  84     790   38.73%
               2922.4515    85      2311.0000      User:  81     901   26.46%
*   0+     0                 0      2403.0000      2922.4515     901   21.62%
               2737.1052    83      2403.0000      User:  83    1021   13.90%
               2680.1757    83      2403.0000      User:  81    1088   11.53%
               2622.5532    84      2403.0000      User:  83    1206    9.14%
*   0+     0                 0      2535.0000      2622.5532    1206    3.45%
                  cutoff            2535.0000      User:  84    1320    0.00%

Mixed integer rounding cuts applied:   15
Gomory fractional cuts applied     :    2
User cuts applied                  : 1739

Number of nodes : 0 (0)
Elapsed CPU time: 3.18 sec.
Cut     CPU time: 0.19 sec. [5.97% of total]
Init LP =  7635.1703  Root LP =  2535  Best IP =  2535  Optim Gap = 0.00%
```

In 21 iterations 1739 (user) cuts (4.5) are added to the formulation and the LP upper bound is improved from 7635.17 to 2535, which is the optimal integer objective value. The lines with * show the improvement of the best integer solution found by CPLEX. The problem is solved in 3 seconds without the need for enumeration. This example illustrates the usefulness of the polyhedral cutting planes in solving robust combinatorial optimization problems.

Appendix

ILOG Concert[4] code that is used to input the robust knapsack model (4.4) to CPLEX solver is listed below. Concert is an object oriented library useful in modeling mathematical programs and building subroutines (such as for cut generation) that communicate with callable libraries of the CPLEX MIP solver. The `CutCallback` function, which is not listed below, implements a simple shortest path algorithm to find violated cuts as described in Section 4.2.

```
int main(int argc, char **argv) {
  IloEnv env;
  try {
     IloInt i, N=100;
// Data
     IloInt b, r;
     IloIntArray a(env,N);
     IloIntArray c(env,N);
     IloIntArray d(env,N);
// Decision variables
     IloNumVarArray x(env,N,0,1,ILOINT);
     IloNumVarArray y(env,N,0,IloInfinity,ILOFLOAT);
     IloNumVar      z(env,0,IloInfinity,ILOFLOAT);
// Initialize data a,b,c,d,r
     .....
// Specify the model
     IloModel model(env);
// Objective function
     IloExpr obj(env);
     obj = IloScalProd(c,x)-IloSum(y)-r*z;
     model.add( IloMinimize(env, obj) );
// Knapsack constraint
     model.add( IloScalProd(a,x) <= b );
// Robustness constraints
    for(i=0; i<N; i++) model.add( z+y[i] >= d[i]*x[i] );
// Cutting plane callback function
     IloCplex cplex(model);
     cplex.use(CutCallback( env,N,d,x,y,z ));
// Solve the problem
     cplex.solve()
// Print out the optimal solution
    env.out()<<''Objective value = ''<<cplex.getObjValue()<<endl;
    for(i=0; i<N; i++){
      env.out()<<"x["<<i<<"]="<<cplex.getValue(x[i])<<endl;
      env.out()<<"y["<<i<<"]="<<cplex.getValue(y[i])<<endl;
    }
    env.out()<<"z="<<cplex.getValue(z)<<endl;
  }
  env.end();
  return 0;
} // end main
```

---

[4]Concert is a trademark of ILOG, Inc.

References

[AS03] A. Atamtürk and M.W.P. Savelsbergh, *Integer programming software systems*, Research Report BCOL.03.01, University of California at Berkeley, January 2003, Revised in October 2003.

[Ata01] A. Atamtürk, *Sequence independent lifting for mixed–integer programming*, Research Report BCOL.01.02, University of California at Berkeley, May 2001, Revised in November 2002. Forthcoming in *Operations Research.*

[Ata03a] ———, *Cover and pack inequalities for (mixed) integer programming*, Research Report BCOL.03.02, University of California at Berkeley, February 2003, Revised in January 2003. Forthcoming in *Annals of Operations Research.*

[Ata03b] ———, *On the facets of the mixed–integer knapsack polyhedron*, Mathematical Programming **98** (2003), 145–175.

[Ata03c] ———, *Strong formulations of robust mixed 0–1 programming*, Research Report BCOL.03.04, University of California at Berkeley, December 2003.

[Bal75a] E. Balas, *Disjunctive programming: Cutting planes from logical conditions*, Nonlinear Programming 2 (O. L. Mangasarian, ed.), Academic Press, 1975, pp. 279–312.

[Bal75b] ———, *Facets of the knapsack polytope*, Mathematical Programming **8** (1975), 146–164.

[Bal79] ———, *Disjunctive programming*, Annals of Discrete Mathematics **5** (1979), 3–51.

[BS03] D. Bertsimas and M. Sim, *Robust discrete optimization and network flows*, Mathematical Programming **98** (2003), 49–71.

[BZ78] E. Balas and E. Zemel, *Facets of the knapsack polytope from minimal covers*, SIAM Journal of Applied Mathematics **34** (1978), 119–148.

[BZ84] ———, *Lifting and complementing yields all facets of positive zero–one programming polytopes*, Proceedings of the International Conference on Mathematical Programming (R. W. Cottle et al., ed.), 1984, pp. 13–24.

[CCMW98] S. Ceria, C. Cordier, H. Marchand, and L. A Wolsey, *Cutting planes for integer programs with general integer variables*, Mathematical Programming **81** (1998), 201–214.

[Chv73] V. Chvátal, *Edmonds polytopes and a hierarchy of combinatorial problems*, Discrete Mathematics **4** (1973), 305–337.

[CKS90] W. Cook, R. Kannan, and A. Schrijver, *Chvátal closures for mixed integer programming problems*, Mathematical Programming **47** (1990), 155–174.

[GJ79] M. R. Garey and D. S. Johnson, *Computers and intractability: A guide to the theory of NP-Completeness*, W. H. Freeman and Company, New York, 1979.

[GLS93] M. Grötschel, L. Lovász, and A. Schrijver, *Geometric algorithms and combinatorial optimization*, Springer-Verlag, 1993.

[GNS98] Z. Gu, G. L. Nemhauser, and M. W. P. Savelsbergh, *Cover inequalities for 0–1 integer programs: Computation*, INFORMS Journal on Computing **10** (1998), 427–437.

[GNS99] ———, *Cover inequalities for 0–1 integer programs: Complexity*, INFORMS Journal on Computing **11** (1999), 117–123.

[GNS00] ———, *Sequence independent lifting in mixed integer programming*, Journal of Combinatorial Optimization **4** (2000), 109–129.

[Gom58] R. E. Gomory, *Outline of an algorithm for integer solutions to linear programming*, Bulletin of the American Mathematical Society **64** (1958), 275–278.

[Gom60a] ———, *An algorithm for the mixed integer problem*, Tech. Report RM-2597, The Rand Corporation, 1960.

[Gom60b] ———, *Solving linear programming problems in integers*, Combinatorial Analysis (R. E. Bellman and M. Hall, Jr., eds.), American Mathematical Society, 1960, pp. 211–216.

[Gom69] ———, *Some polyhedra related to combinatorial problems*, Linear Algebra and Its Applications **2** (1969), 451–558.

[HJP75] P. L. Hammer, E. L. Johnson, and U. N. Peled, *Facets of regular 0–1 polytopes*, Mathematical Programming **8** (1975), 179–206.

[JP81] E. L. Johnson and M. W. Padberg, *A note on knapsack problem with special ordered sets*, Operations Research Letters **1** (1981), 18–22.

[Kha79] L. G. Khachian, *A polynomial algorithm for linear programming*, Soviet Mathematics Doklady **20** (1979), 191–194.

[KNT98] D. Klabjan, G. L. Nemhauser, and C. Tovey, *The complexity of cover inequality separation*, Operations Research Letters **23** (1998), 35–40.

[KP80] R. M. Karp and C. H. Papadimitriou, *On the linear characterizations of combinatorial optimization problems*, Proceedings of the 21st Annual Symposium on Foundations of Computer Science, 1980, pp. 1–9.

[Las01] J. B. Lasserre, *An explicit exact SDP relaxation for nonlinear 0-1 programs*, Lecture notes in Computer Science (K. Aardal and A. M. H. Gerards, eds.), vol. 2081, 2001, pp. 293–303.

[LS91] L. Lovász and A. Schrijver, *Cones of matrices and set-functions and 0-1 optimization*, SIAM Journal on Optimization **1** (1991), 166–190.

[LW03] Q. Louveaux and L. A. Wolsey, *Lifting, superadditivity, mixed integer rounding, and single node flow sets revisited*, CORE Discussion Paper 2003/1, Université Catholique de Louvain, 2003.

[Mey74] R. R. Meyer, *On the existence of optimal solutions to integer and mixed-integer programming problems*, Mathematical Programming **7** (1974), 223–235.

[MW99] H. Marchand and L. A. Wolsey, *The 0-1 knapsack problem with a single continuous variable*, Mathematical Programming **85** (1999), 15–33.

[MW01] ———, *Aggregation and mixed integer rounding to solve MIPs*, Operations Research **49** (2001), 363–371.

[NW88] G. L. Nemhauser and L. A. Wolsey, *Integer and combinatorial optimization*, John Wiley and Sons, New York, 1988.

[NW90] ———, *A recursive procedure for generating all cuts for 0–1 mixed integer programs*, Mathematical Programming **46** (1990), 379–390.

[Pad73] M. W. Padberg, *On the facial structure of set packing polyhedra*, Mathematical Programming **5** (1973), 199–215.

[Pad79] ———, *A note on 0–1 programming*, Operations Research **23** (1979), 833–837.

[RdN02] J.-P. P. Richard, I. R. de Farias, and G. L. Nemhauser, *Lifted inequalities for 0–1 mixed integer programming : Basic theory and algorithms*, Integer Programming and Combinatorial Optimization, 9th International IPCO Conference, Proceedings, 2002, pp. 161–175.

[Roc70] R. T. Rockafellar, *Convex analysis*, Princeton University Press, Princeton, New Jersey, 1970.

[SA90] H. D. Sherali and W. P. Adams, *Hierarchy of relaxations between the continuous and convex hull representations for zero-one programming problems*, SIAM Journal on Discrete Mathematics **3** (1990), 411–430.

[Sch87] A. Schrijver, *Theory of linear and integer programming*, John Wiley and Sons, Chichester, 1987.

[Sch03] ———, *Combinatorial optimization: Polyhedra and efficiency*, Springer Verlag, Berlin, 2003.

[Wol75] L. A. Wolsey, *Faces for linear inequality in 0-1 variables*, Mathematical Programming **8** (1975), 165–178.

[Wol76] ———, *Facets and strong valid inequalities for integer programs*, Operations Research **24** (1976), 367–372.

[Wol77] ———, *Valid inequalities and superadditivity for 0/1 integer programs*, Mathematics of Operations Research **2** (1977), 66–77.

[Wol98] ———, *Integer programming*, John Wiley and Sons, New York, 1998.

[Zem78] E. Zemel, *Lifting the facets of zero-one polytopes*, Mathematical Programming **15** (1978), 268–277.

[Zem89] ———, *Easily computable facets of the knapsack polytope*, Mathematics of Operations Research **14** (1989), 760–764.

Department of Industrial Engineering and Operations Research, University of California, Berkeley, California 94720–1777, USA.

E-mail address: `atamturk@ieor.berkeley.edu`

Proceedings of Symposia in Applied Mathematics
Volume **61**, 2004

Graphs and Combinatorial Optimization

Gérard Cornuéjols

Abstract. The integer programming models known as set packing and set covering have a wide range of applications, many of which arise in the context of graph theory. Sometimes, because of the special structure of the constraint matrix, the natural linear programming relaxation yields an optimal solution that is integral, thus solving the problem. Sometimes, both the linear programming relaxation and its dual have integer optimal solutions. Under which conditions do such integrality properties hold? This question is of both theoretical and practical interest. Min-max theorems, polyhedral combinatorics and graph theory all come together in this rich area of discrete mathematics. In addition to min-max and polyhedral results, some of the deepest results in this area come in two flavors: "excluded structure" characterizations and "decomposition" results. These notes provide an introduction to this area. In particular, they survey the celebrated Strong Perfect Graph Conjecture and its recent solution by Chudnovsky, Robertson, Seymour and Thomas, and Lehman's characterization of ideal clutters. The monograph "Combinatorial Optimization: Packing and Covering" by Cornuéjols, CBMS 74, SIAM (2001) covers related material.

1. Clutters

A *clutter* $\mathcal{C}$ is a family $E(\mathcal{C})$ of subsets of a finite ground set $V(\mathcal{C})$ with the property that $S_1 \not\subseteq S_2$ for all distinct $S_1, S_2 \in E(\mathcal{C})$. $V(\mathcal{C})$ is called the set of *vertices* and $E(\mathcal{C})$ the set of *edges* of $\mathcal{C}$. Clutters are also called *Sperner families* in the literature. A clutter is *trivial* if it has no edge or if it has the empty set as the only edge.

For example, a simple graph (no multiple edges or loops) is a clutter where every edge has cardinality two. We refer the reader to West [**56**] for a basic reference in graph theory. In these notes, all graphs are assumed to be finite and simple.

In a clutter, a *matching* is a set of pairwise disjoint edges. A *transversal* is a set of vertices that intersects all the edges. A clutter is said to *pack* if the maximum cardinality of a matching equals the minimum cardinality of a transversal. This terminology is due to Seymour [**54**]. Many min-max theorems in graph theory can be rephrased by saying that a clutter packs. We give three examples. The first is König's theorem (for a proof, see Theorem 3.1.11 in West [**56**]).

1991 *Mathematics Subject Classification.* 05C17, 90C27.

Key words and phrases. Combinatorial optimization, packing, covering, ideal clutter, perfect graph.

This work was supported in part by NSF Grant DMI-0098427 and ONR grant N00014-97-1-0196.

THEOREM 1.1. (König [**37**]) *In a bipartite graph, the maximum cardinality of a matching equals the minimum cardinality of a transversal.*

As a second example, consider the edge version of Menger's theorem (for a proof, see Theorem 4.2.18 in West [**56**]).

THEOREM 1.2. (Menger [**44**]) *Let s and t be distinct nodes of a graph G. The maximum number of pairwise edge-disjoint st-paths in G equals the minimum number of edges in an st-cut.*

Let $\mathcal{C}_1$ be the clutter whose vertices are the edges of G and whose edges are the st-paths of G (Following West's terminology [**56**], paths and cycles have no repeated nodes). We call $\mathcal{C}_1$ the *clutter of st-paths.* Its transversals are the st-cuts. Thus, rephrased in terms of clutters, Menger's theorem states that the clutter of st-paths packs.

Interestingly, some famous results and difficult conjectures can be rephrased by saying that certain clutters pack. In a graph G, a set of edges is called a *postman set* if it induces an acyclic graph whose odd degree nodes coincide with the odd degree nodes of G.

CONJECTURE 1.3. (Conforti and Johnson [**19**]) *The clutter of postman sets packs in planar graphs.*

It turns out that, if true, this conjecture implies the four color theorem [**1**] stating that every planar graph is 4-colorable, i.e. the nodes of a planar graph can be colored with four colors so that no edge has its endnodes colored with the same color (see [**20**] for details). This indicates that a full understanding of the clutters that pack must be extremely difficult. More restricted notions are amenable to beautiful theories, while still containing rich classes of examples. We introduce several such concepts in the following section.

1.1. The Max Flow Min Cut Property and Idealness. Given a nontrivial clutter $\mathcal{C}$, we define $M(\mathcal{C})$ to be a 0,1 matrix whose columns are indexed by $V(\mathcal{C})$, whose rows are indexed by $E(\mathcal{C})$ and where $m_{ij} = 1$ if and only if the vertex corresponding to column j belongs to the edge corresponding to row i. In other words, the rows of $M(\mathcal{C})$ are the characteristic vectors of the sets in $E(\mathcal{C})$. Note that the definition of $M(\mathcal{C})$ is unique up to permutation of rows and permutation of columns. Furthermore, $M(\mathcal{C})$ contains no dominating row, since $\mathcal{C}$ is a clutter (A vector $r \in F$ is said to be *dominating* if there exists $v \in F$ distinct from r such that $r \geq v$). A 0,1 matrix containing no dominating rows is called a *clutter matrix.* Given any 0,1 clutter matrix M, let $\mathcal{C}(M)$ denote the clutter such that $M(\mathcal{C}(M)) = M$.

Let $M \neq 0$ be a 0,1 clutter matrix and consider the following pair of dual linear programs, where x and y denote the vectors of unknowns.

$$\min\{wx : x \geq 0,\ Mx \geq \mathbf{1}\} \tag{1.1}$$

$$= \max\{y\mathbf{1} : y \geq 0,\ yM \leq w\} \tag{1.2}$$

Here x and $\mathbf{1}$ are column vectors while w and y are row vectors. $\mathbf{1}$ denotes a vector all of whose components are equal to 1. Equality holds between (1.1) and (1.2) as a consequence of the duality theorem of linear programming (see for example Section 7.4 in Schrijver [**52**]).

We say that a vector is *integral* if all its components are integers. In these notes, our interest will focus on integral solutions to (1.1) or (1.2). First, we reformulate the definition of a clutter that packs.

DEFINITION 1.4. Clutter $\mathcal{C}(M)$ *packs* if both (1.1) and (1.2) have optimal solution vectors x and y that are integral when $w = \mathbf{1}$.

This corresponds to the earlier definition because, when $w = \mathbf{1}$, if (1.1) and (1.2) have integral optimal solutions, they also have 0,1 optimal solutions x and y. Furthermore, x is the characteristic vector of a transversal of $\mathcal{C}(M)$ and y is the characteristic vector of a matching.

DEFINITION 1.5. Clutter $\mathcal{C}(M)$ has the *packing property* if both (1.1) and (1.2) have optimal solution vectors x and y that are integral for all vectors w with components equal to $0, 1$ or $+\infty$. (Note: When $w_j = 0$, it is optimal to set $x_j = 1$ and when $w_j = +\infty$, it is optimal to set $x_j = 0$. By convention, we consider that $w_j x_j = 0$ in this case.)

DEFINITION 1.6. Clutter $\mathcal{C}(M)$ has the *Max Flow Min Cut* property (or MFMC property) if both (1.1) and (1.2) have optimal solution vectors x and y that are integral for all nonnegative integral vectors w.

Clearly, the MFMC property for a clutter implies the packing property which itself implies that the clutter packs. Conforti and Cornuéjols [**15**] conjectured that, in fact, the MFMC property and the packing property are identical. This conjecture is still open.

CONJECTURE 1.7. *A clutter has the MFMC property if and only if it has the packing property.*

DEFINITION 1.8. Clutter $\mathcal{C}(M)$ is *ideal* if (1.1) has an optimal solution vector x that is integral for all $w \geq 0$.

It is easy to show that the MFMC property implies idealness. Indeed, if (1.1) has an optimal solution vector x for all nonnegative integral vectors w, then (1.1) has an optimal solution x for all nonnegative rational vectors w. In fact, the packing property implies idealness.

THEOREM 1.9. *If a clutter has the packing property, then it is ideal.*

This follows from a result of Lehman [**39**] (see Theorem 3.19 presented later).

Figure 1 shows inclusion relationships between the classes of clutters introduced above. The following clutter matrices show that some of these inclusions are strict:

$$Q_6 = \begin{pmatrix} 1 & 1 & 0 & 1 & 0 & 0 \\ 1 & 0 & 1 & 0 & 1 & 0 \\ 0 & 1 & 1 & 0 & 0 & 1 \\ 0 & 0 & 0 & 1 & 1 & 1 \end{pmatrix}, \; C_3^2 = \begin{pmatrix} 1 & 1 & 0 \\ 0 & 1 & 1 \\ 1 & 0 & 1 \end{pmatrix}, \; C_4^2 = \begin{pmatrix} 1 & 1 & 0 & 0 \\ 0 & 1 & 1 & 0 \\ 0 & 0 & 1 & 1 \\ 1 & 0 & 0 & 1 \end{pmatrix}, \; Q_6^+$$

and C_3^{2+}: For an $m \times n$ 0,1 matrix M, we use M^+ to denote the $m \times (n+1)$ matrix obtained from M by appending the column vector $\mathbf{1}$. The reader is encouraged to verify that these five clutters satisfy the properties suggested in Figure 1.

A subset P of $\mathbb{R}^n$ is called a *polyhedron* if it is the intersection of finitely many affine half-spaces, i.e. $P = \{x : Ax \leq b\}$ for some matrix A and vector b. Observe that $\mathcal{C}(M)$ is ideal if and only if the polyhedron $P = \{x \geq 0 : Mx \geq \mathbf{1}\}$ is an *integral*

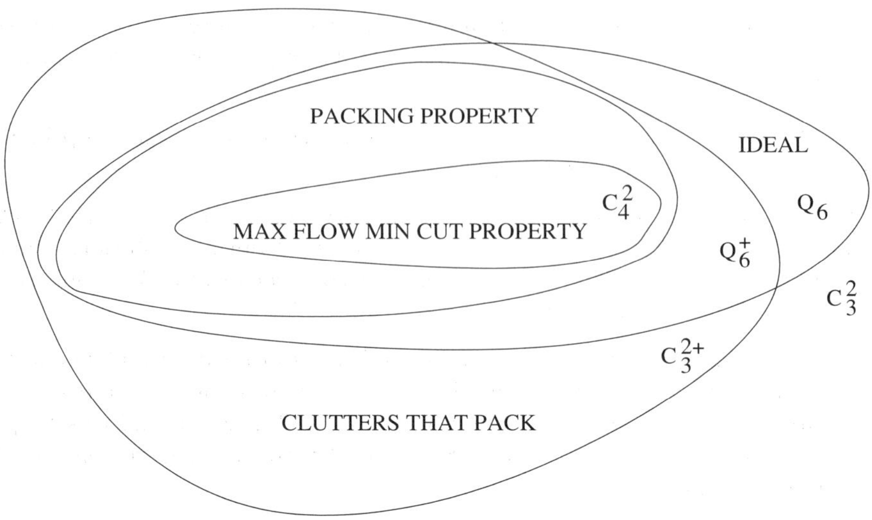

FIGURE 1. Classes of clutters.

polyhedron, that is, all the extreme points of P have only integer coordinates (an *extreme point* of P is a point x that cannot be written as a convex combination of $x^1, x^2 \in P$ where $x^1 \neq x$ and $x^2 \neq x$). Equivalently, $\mathcal{C}$ is ideal if and only if

$$\begin{aligned} x(S) &\geq 1 \quad \text{for all } S \in E(\mathcal{C}) \\ x &\geq 0 \end{aligned}$$

is an integral polyhedron, where $x(S)$ denotes $\sum_{i \in S} x_i$.

A linear system $Ax \geq b$ is *Totally Dual Integral* (TDI) if the linear program $\min\{wx : Ax \geq b\}$ has an integral optimal dual solution y for every integral w for which this linear program has a finite optimum. Edmonds and Giles [**27**] showed that, if $Ax \geq b$ is TDI and b is integral, then $P = \{x : Ax \geq b\}$ is an integral polyhedron. The interested reader can find the proof of the Edmonds-Giles theorem in Schrijver [**52**] pages 310–311, or Nemhauser and Wolsey [**45**] pages 536–537. It follows that $\mathcal{C}(M)$ has the MFMC property if and only if (1.2) has an optimal integral solution y for all nonnegative integral vectors w.

For convenience, we say that trivial clutters have all the above properties: MFMC, idealness, etc.

1.2. Perfection. The min-max equation (1.1)=(1.2) has a close max-min relative

$$\max\{wx : x \geq 0,\ Mx \leq \mathbf{1}\} \tag{1.3}$$

$$= \min\{y\mathbf{1} : y \geq 0,\ yM \geq w\} \tag{1.4}$$

The clutter $\mathcal{C}(M)$ *covers* if both (1.3) and (1.4) have optimal solution vectors x and y that are integral when $w = \mathbf{1}$. This is the analog of a clutter that packs. Like for clutters that pack, a full understanding of the clutters that cover appears to be extremely difficult. Thus like for the min-max equation (1.1)=(1.2) where we introduced the notions of packing property, MFMC property and idealness, it makes sense to introduce three analogous definitions for the max-min equation (1.3)=(1.4). Surprisingly, these three notions turn out to be identical.

DEFINITION 1.10. A *perfect matrix* M is a $0,1$ matrix with no column of 0's such that $P = \{x \geq 0 : \ Mx \leq \mathbf{1}\}$ is an integral polyhedron.

When M is a perfect matrix, the linear program $\max\{wx : \ x \geq 0, \ Mx \leq \mathbf{1}\}$ has an integral optimal solution x for all w, and therefore the *set packing problem* $\max\{wx : \ Mx \leq \mathbf{1}, \ x \in \{0,1\}^n\}$ can be solved in polynomial time as a linear program. By contrast, for a general 0,1 matrix M, the set packing problem is NP-hard [**32**].

As we mentioned already, Edmonds and Giles [**27**] observed that, when a linear system $Ax \leq b, \ x \geq 0$ is TDI and b is integral, the polyhedron $\{x : \ Ax \leq b, \ x \geq 0\}$ is integral. The converse is not true in general (recall Q_6 in Figure 1 for the linear system $-Q_6 x \leq -\mathbf{1}, \ x \geq 0$). But it is true when A is a 0,1 matrix and $b = \mathbf{1}$.

THEOREM 1.11. (Lovász [**40**], Fulkerson [**31**], Chvátal [**10**]) *For a 0,1 matrix M with no column of 0's, the following statements are equivalent:*

(i) *the linear system $Mx \leq \mathbf{1}, \ x \geq 0$ is TDI,*
(ii) *the matrix M is perfect,*
(iii) *max $\{wx : \ Mx \leq \mathbf{1}, \ x \geq 0\}$ has an integral optimal solution x for all $w \in \{0,1\}^n$.*

Clearly (i) implies (ii) implies (iii), where the first implication is the Edmonds-Giles property and the second follows from the definition of perfection. What is surprising is that (iii) implies (i) and, in fact, that (ii) implies (i). We will prove this theorem in the next section. It is convenient to present this material in terms of graphs.

The concept of perfect graph (to be defined in the next section) was introduced by Berge [**2**] at a workshop in 1960. In a graph, a *clique* is a set of pairwise adjacent nodes. The *clique-node matrix* of a graph is a 0,1 matrix M with columns indexed by the nodes of G where the rows of M are the characteristic vectors of the maximal cliques of G. Thus entry m_{ij} of M is 1 if and only if node j belongs to maximal clique i. Chvátal [**10**] established the following connection between perfect graphs and perfect matrices: A 0,1 clutter matrix with no column of 0's is perfect if and only if it is the clique-node matrix of a perfect graph. Therefore we may reason in terms of graphs rather than 0,1 matrices.

2. Perfect Graphs

The node set of graph G is denoted by $V(G)$ and its edge set by $E(G)$. A *stable set* is a set of nodes no two of which are adjacent. A *clique* is a set of nodes every pair of which are adjacent. The cardinality of a largest clique in graph G is denoted by $\omega(G)$. The cardinality of a largest stable set is denoted by $\alpha(G)$. A *k-coloring* is a partition of the nodes into k stable sets (these stable sets are called *color classes*). The *chromatic number* $\chi(G)$ is the smallest value of k for which there exists a k-coloring. Obviously, $\omega(G) \leq \chi(G)$ since the nodes of a clique must be in distinct color classes of a k-coloring. An *induced subgraph* of G is a graph with node set $S \subseteq V(G)$ and edge set comprising all the edges of G with both ends in S. It is denoted by $G(S)$. The graph $G(V(G) - S)$ is denoted by $G \setminus S$. A graph G is *perfect* if $\omega(H) = \chi(H)$ for every induced subgraph H of G. A graph is *minimally imperfect* if it is not perfect but all its proper induced subgraphs are.

A *hole* is a graph induced by a cycle of length at least 4. A hole is *odd* if it contains an odd number of nodes. Odd holes are not perfect since their chromatic number is 3 whereas the size of their largest clique is 2. It is easy to check that odd holes are minimally imperfect. The complement of a graph G is the graph $\bar{G}$ with the same node set as G, and uv is an edge of $\bar{G}$ if and only if it is not an edge of G. It is easy to check that complements of odd holes are also minimally imperfect. In the early sixties Berge [**2**] proposed the *Strong Perfect Graph Conjecture*: The odd holes and their complements are the only minimally imperfect graphs. This conjecture attracted much attention over the next forty years. It was proved in May 2002 by Chudnovsky, Robertson, Seymour and Thomas [**7**] in a very impressive paper. Claude Berge passed away in June 2002 knowing that his famous conjecture is true.

THEOREM 2.1. **(Strong Perfect Graph Theorem)** (Chudnovsky, Robertson, Seymour and Thomas [**7**]) *The only minimally imperfect graphs are the odd holes and their complements.*

In this section, we survey key aspects of the proof of the Strong Perfect Graph Theorem. A *Berge graph* is a graph that does not contain an odd hole or its complement as an induced subgraph. Clearly, every perfect graph is a Berge graph. The Strong Perfect Graph Theorem states that the converse is also true: Every Berge graph is perfect. The idea of the proof is to show that every Berge graph either falls into one of four basic classes of perfect graphs, or that it has a kind of separation that cannot occur in a minimally imperfect graph. This implies that there exists no minimally imperfect Berge graph, and therefore every Berge graph is perfect.

In [**2**], Berge also made a weaker conjecture, which states that a graph G is perfect if and only if its complement $\bar{G}$ is perfect. This conjecture was proved by Lovász [**40**] in 1972. We give a short elegant proof due to Gasparyan [**33**] by proving first the following stronger result.

THEOREM 2.2. (Lovász [**41**]) *A graph G is perfect if and only if, for every induced subgraph H, the number of nodes of H is at most $\alpha(H)\omega(H)$.*

PROOF. First assume that G is perfect. Then, for every induced subgraph H, $\omega(H) = \chi(H)$. Since the number of nodes of H is at most $\alpha(H)\chi(H)$, the inequality follows.

We give a proof of the converse due to Gasparyan [**33**]. Assume that G is not perfect. Let H be a minimally imperfect subgraph of G and let n be the number of nodes of H. Let $\alpha = \alpha(H)$ and $\omega = \omega(H)$. Then H satisfies

$$\omega = \chi(H \backslash v) \text{ for every node } v \in V(H)$$

$$\text{and } \omega = \omega(H \backslash S) \text{ for every stable set } S \subseteq V(H).$$

Let A_0 be a stable set of size α in H. Fix an ω-coloring of each of the α graphs $H \backslash s$ for $s \in A_0$, let $A_1, \ldots, A_{\alpha\omega}$ be the stable sets occuring as a color class in one of these colorings and let $\mathcal{A} := \{A_0, A_1, \ldots, A_{\alpha\omega}\}$. Let $\mathbf{A}$ be the corresponding stable set versus node incidence matrix. Define $\mathcal{B} := \{B_0, B_1, \ldots, B_{\alpha\omega}\}$ where B_i is a clique of size ω in $H \backslash A_i$. Let $\mathbf{B}$ be the corresponding clique versus node incidence matrix.

Claim: Every clique of size ω in H intersects all but one of the stables sets in $\mathcal{A}$.

Proof: Let $S_1, \ldots, S_\omega$ be any ω-coloring of $H \setminus v$. Since any clique C of size ω in H has at most one node in each S_i, C intersects all S_i's if $v \notin C$ and all but one if $v \in C$. Since C has at most one node in A_0, the claim follows.

In particular, it follows that $\mathbf{AB}^T = J - I$ where J is the matrix filled with ones and I the identity. Since $J - I$ is nonsingular, $\mathbf{A}$ and $\mathbf{B}$ have at least as many columns as rows, that is $n \geq \alpha\omega + 1$. □

COROLLARY 2.3. **(Perfect Graph Theorem)** (Lovász [**40**]) *Graph G is perfect if and only if graph $\bar{G}$ is perfect.*

PROOF. Since, for any graph H, $\alpha(H) = \omega(\bar{H})$ and $\omega(H) = \alpha(\bar{H})$, Theorem 2.2 implies Corollary 2.3.

□

2.1. Four Basic Classes of Perfect Graphs. Bipartite graphs are perfect since, for any induced subgraph H, the bipartition implies that $\chi(H) \leq 2$ and therefore $\omega(H) = \chi(H)$.

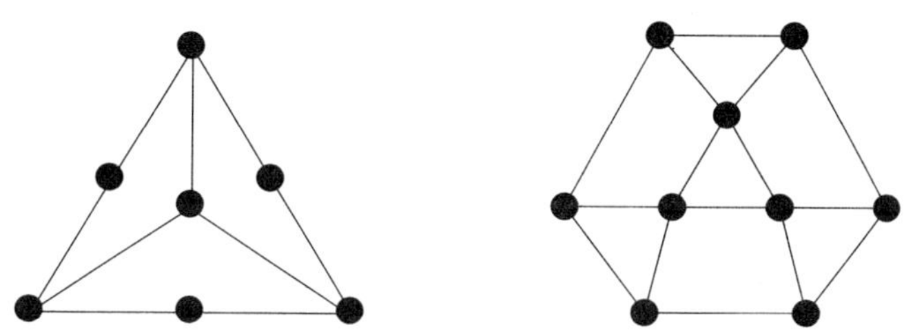

FIGURE 2. A bipartite graph and its line graph.

A graph L is the *line graph* of a graph G if $V(L) = E(G)$ and two nodes of L are adjacent if and only if the corresponding edges of G share a vertex. See Figure 2 for an example.

PROPOSITION 2.4. *Line graphs of bipartite graphs are perfect.*

PROOF. If G is bipartite, $\chi'(G) = \Delta(G)$ by a theorem of König [**37**], where χ' denotes the edge-chromatic number and Δ the largest node degree.

If L is the line graph of a bipartite graph G, then $\chi(L) = \chi'(G)$ and $\omega(L) = \Delta(G)$. Therefore $\chi(L) = \omega(L)$. Since induced subgraphs of L are also line graphs of bipartite graphs, the result follows. □

Since bipartite graphs and line graphs of bipartite graphs are perfect, it follows from Lovász's perfect graph theorem (Corollary 2.3) that the complements of bipartite graphs and of line graphs of bipartite graphs are perfect. This can also be verified directly, without using the perfect graph theorem. To summarize, in this section we have introduced four classes of perfect graphs:

- bipartite graphs and their complements, and
- line graphs of bipartite graphs and their complements.

These graphs are called *basic.*

2.2. 2-Join, Homogeneous Pair and Skew Partition. The proof of the Strong Perfect Graph Theorem by Chudnovsky, Robertson, Seymour and Thomas is based on a decomposition theorem for Berge graphs: Every Berge graph is basic as defined above, or it can be decomposed using a 2-join, a homogeneous pair or a skew partition. We define these three decompositions in this section.

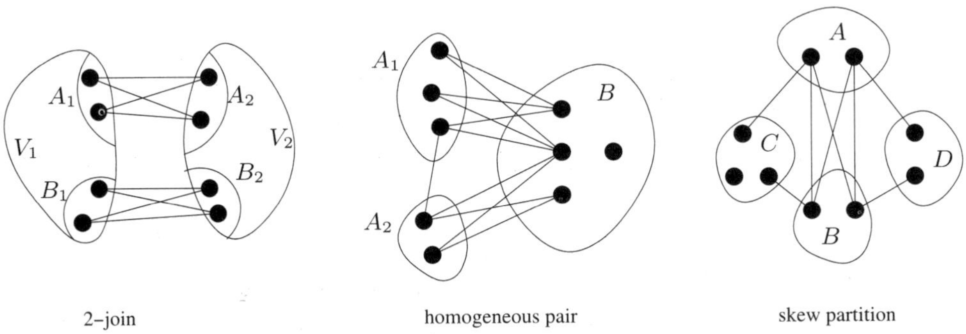

FIGURE 3. Decompositions.

2-Join

A graph G has a *2-join* if its nodes can be partitioned into sets V_1 and V_2, each of cardinality at least three, with nonempty disjoint subsets $A_1, B_1 \subseteq V_1$ and $A_2, B_2 \subseteq V_2$, such that all the nodes of A_1 are adjacent to all the nodes of A_2, all the nodes of B_1 are adjacent to all the nodes of B_2 and these are the only adjacencies between V_1 and V_2. See Figure 3. 2-joins were introduced by Cornuéjols and Cunningham [**21**] in 1985. They gave an $O(|V(G)|^2|E(G)|^2)$ algorithm to find whether a graph G has a 2-join.

When G contains a 2-join, we can decompose G into two blocks G_1 and G_2 defined as follows.

DEFINITION 2.5. If A_2 and B_2 are in different connected components of $G(V_2)$, define *block* G_1 to be $G(V_1 \cup \{p_1, q_1\})$, where $p_1 \in A_2$ and $q_1 \in B_2$. Otherwise, let P_1 be a shortest path from A_2 to B_2 in $G(V_2)$ and define *block* G_1 to be $G(V_1 \cup V(P_1))$. Block G_2 is defined similarly.

THEOREM 2.6. **(2-Join Decomposition Theorem)** (Cornuéjols and Cunningham [**21**], see also [**20**]) *Graph G is perfect if and only if its blocks G_1 and G_2 are perfect.*

COROLLARY 2.7. *If a minimally imperfect graph G has a 2-join, then G is an odd hole.*

PROOF. Since G is not perfect, Theorem 2.6 implies that block G_1 or G_2 is not perfect, say G_1. Since G_1 is an induced subgraph of G and G is minimally imperfect, it follows that $G = G_1$. Thus, since $|V_2| \geq 3$, V_2 induces a chordless path P_1. Therefore G is a minimally imperfect graph with a node of degree 2. It is well known that such a graph G is an odd hole [**47**]. □

Homogeneous Pair

The notion of homogeneous pair was introduced by Chvátal and Sbihi [**13**]. A graph G has a *homogeneous pair* if $V(G)$ can be partitioned into subsets A_1, A_2 and B, such that:

- $|A_1| + |A_2| \geq 3$ and $|B| \geq 2$.
- If a node of B is adjacent to a node of A_i then it is adjacent to all the nodes of A_i, for $i \in \{1, 2\}$.

THEOREM 2.8. **(Homogeneous Pair Theorem)** (Chvátal and Sbihi [**13**]) *No minimally imperfect graph has a homogeneous pair.*

Skew Partition

A graph G has a *skew partition* if its nodes can be partitioned into four nonempty sets A, B, C, D such that there are all the possible edges between A and B and no edges from C to D. Chvátal [**11**] introduced skew partitions in 1985 and he conjectured that no minimally imperfect graph has a skew partition. He observed that the conjecture holds for a *star cutset*, defined to be a skew partition where $|A| = 1$.

LEMMA 2.9. **(Star Cutset Lemma)** (Chvátal [**11**]) *No minimally imperfect graph has a star cutset.*

PROOF. Assume that G is minimally imperfect and has a star cutset. Let G_1 be the graph induced by $A \cup B \cup C$ and G_2 the graph induced by $A \cup B \cup D$. The graphs G_1 and G_2 are perfect. Let S_i be the color class of an $\omega(G)$-coloring of G_i that contains the unique node of A, for $i \in \{1, 2\}$. Then S_i meets all the cliques of size $\omega(G)$ in G_i, i.e. $\omega(G \setminus (S_1 \cup S_2)) < \omega(G)$. It follows that $G \setminus (S_1 \cup S_2)$ can be colored with fewer than $\omega(G)$ colors, since it is perfect. Since $S_1 \cup S_2$ is a stable set, G can be colored with $\omega(G)$ colors, a contradiction. □

Noteworthy contributions towards the skew partition conjecture were made by Hoàng [**36**] and Roussel and Rubio [**51**]. The conjecture was settled by Chudnovsky, Robertson, Seymour and Thomas [**7**]. They obtained it as a consequence of the Strong Perfect Graph Theorem.

THEOREM 2.10. **(Skew Partition Theorem)** (Chudnovsky, Robertson, Seymour and Thomas [**7**]) *No minimally imperfect graph has a skew partition.*

In order to prove the Strong Perfect Graph Theorem, Chudnovsky, Robertson, Seymour and Thomas first proved the following weaker result.

A skew partition is *balanced* if

(i) every induced path of length at least 2 in G with ends in $A \cup B$ and interior in $C \cup D$ is even, and

(ii) every induced path of length at least 2 in $\bar{G}$ with ends in $C \cup D$ and interior in $A \cup B$ is even.

THEOREM 2.11. (Chudnovsky, Robertson, Seymour and Thomas [**6**]) *A minimally imperfect Berge graph with smallest number of nodes cannot have a balanced skew partition.*

We give the proof of Theorem 2.11. It uses Lovász's Replication Lemma [**40**] which we discuss next. Incidentally, the Replication Lemma was the step that Fulkerson missed in his attempt to prove the Perfect Graph Theorem. Because

Fulkerson had convinced himself that the Replication Lemma was likely to be false, he had not tried very hard to prove it. Fulkerson [**31**] says: "In the Spring of 1971, I received a postcard from Berge saying that he had just heard that Lovász had a proof of the perfect graph conjecture. This immediately rekindled my interest, naturally, and so I sat down at my desk and thought again about the replication lemma. Some four or five hours later, I saw a simple proof of it."

LEMMA 2.12. **(Replication Lemma)** (Lovász [**40**]) *Let G be a perfect graph and $v \in V(G)$. Create a new node v' and join it to v and to all the neighbors of v. Then the resulting graph G' is perfect.*

PROOF. It suffices to show $\chi(G') = \omega(G')$ since, for induced subgraphs, the proof follows similarly. We distinguish two cases.

Case 1: Node v is contained in some clique of size $\omega(G)$ in G. Then $\omega(G') = \omega(G) + 1$. This implies $\chi(G') \leq \omega(G')$, since at most one new color is needed in G'. Clearly $\chi(G') = \omega(G')$ follows.

Case 2: Node v is not contained in any clique of size $\omega(G)$ in G. Consider any $\omega(G)$-coloring of G and let S be the color class containing v. Then $\omega(G \setminus (S - \{v\})) = \omega(G) - 1$, since every clique of size $\omega(G)$ in G meets $S - \{v\}$. By the perfection of G, the graph $G \setminus (S - \{v\})$ can be colored with $\omega(G) - 1$ colors. Using one additional color for the nodes $(S - \{v\}) \cup \{v'\}$, we obtain a $\omega(G)$-coloring of G'.

□

Proof of Theorem 2.11: Let G be a minimally imperfect Berge graph with smallest number of nodes. Suppose that G has a balanced skew partition A, B, C, D. Note that each of C, D has cardinality at least two since, otherwise, the minimally imperfect graph $\bar{G}$ has a star cutset, contradicting Lemma 2.9. Let G' be the graph obtained from G by adding a node v adjacent to all the nodes of A and to no other node of G. If G' contains an odd hole, then G has an odd path contradicting (i) in the definition of a balanced skew partition. Similarly, if $\bar{G}'$ contains an odd hole, (ii) is contradicted. Therefore G' is a Berge graph. Now consider $G_1 = G' \setminus D$ and $G_2 = G' \setminus C$. For $i \in \{1, 2\}$, the graph G_i is perfect since it is Berge and has fewer nodes than G. Replicate node v in G_i so that v belongs to a clique of size $\omega(G)$. By the Replication Lemma 2.12, the resulting graph R_i is perfect. Consider $\omega(G)$-colorings of R_1 and R_2 respectively. Both colorings have the same number of colors in A and assume w.l.o.g. that these colors are $1, 2, \ldots, k$. Let K be the subgraph of G induced by the nodes with colors $1, 2, \ldots, k$ and let H be the subgraph of G induced by the nodes with other colors. Since every clique of size $\omega(G)$ in G is in $G \setminus D$ or $G \setminus C$, the largest clique in K has size k and the largest clique in H has size $\omega(G) - k$. The graphs H and K are perfect since they are proper subgraphs of G. Color K with k colors and H with $\omega(G) - k$ colors. Now G is colored with $\omega(G)$ colors, a contradiction to the assumption that G is minimally imperfect. □

Theorem 2.11 was presented in September 2001 at a workshop in Princeton. As the next step towards Theorem 2.10, Chudnovsky and Seymour obtained the following theorem in January 2002. Its proof is much more difficult than that of Theorem 2.11.

THEOREM 2.13. (Chudnovsky and Seymour [**8**]) *A minimally imperfect Berge graph with smallest number of nodes cannot have a skew partition.*

2.3. Decomposition of Berge Graphs. Conforti, Cornuéjols and Vušković proposed the following approach to solving the Strong Perfect Graph Conjecture.

CONJECTURE 2.14. (Conforti, Cornuéjols and Vušković (2001)) **(Decomposition Conjecture)** *Every Berge graph G is basic or has a skew partition, or one of G or $\bar{G}$ has a 2-join.*

Chudnovsky, Robertson, Seymour and Thomas proved the following variation of this conjecture.

THEOREM 2.15. (Chudnovsky, Robertson, Seymour and Thomas [**7**]) **(Decomposition Theorem)** *Every Berge graph G is basic or has a skew partition or a homogeneous pair, or one of G or $\bar{G}$ has a 2-join.*

This theorem implies the Strong Perfect Graph Theorem. Indeed, suppose that the Decomposition Theorem holds and that there exists a minimally imperfect graph G distinct from an odd hole or its complement. Choose G with the smallest number of nodes. G cannot have a skew partition by Theorem 2.13. G cannot have a homogeneous pair by Theorem 2.8. Neither G nor $\bar{G}$ can have a 2-join by Corollary 2.7. Since G is a Berge graph, G must be basic by the Decomposition Theorem. Therefore G is perfect, a contradiction.

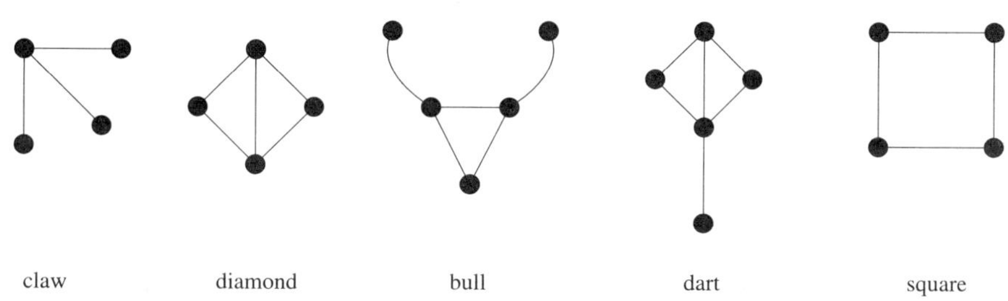

FIGURE 4. Small graphs.

Theorem 2.15 was already known to hold in several special cases. For example, in 1984 Burlet and Fonlupt [**4**] worked out the case when G is a Meyniel graph (A graph is *Meyniel* if all its odd cycles have at least two chords). For a given graph X, a graph G is *X-free* if G does not contain X as an induced subgraph. Theorem 2.15 was known to hold for several classes of X-free graphs (see Figure 4). For example, it was known when G is claw-free (Chvatal and Sbihi [**14**] in 1988 and Maffray and Reed [**43**] in 1999), diamond-free (Fonlupt and Zemirline [**29**] in 1987), bull-free (Chvátal and Sbihi [**13**] in 1987), or dart-free (Chvátal, Fonlupt, Sun and Zemirline [**12**] in 2000). All these results involve special types of skew partitions (such as star cutsets) and, in some cases, homogeneous pairs [**13**]. A special case of 2-join called augmentation of a flat edge appears in [**43**]. In 1999, Conforti and Cornuéjols [**16**] used more general 2-joins to prove Conjecture 2.14 for WP-free Berge graphs, a class of graphs that contains all bipartite graphs and all line graphs of bipartite graphs. [**16**] was the precursor of a sequence of decomposition results involving 2-joins. The following result was obtained in February 2001. A *square* is a hole of length four.

THEOREM 2.16. (Conforti, Cornuéjols and Vušković [**17**]) *A square-free Berge graph is bipartite, the line graph of a bipartite graph, or has a 2-join or a star cutset.*

A breakthrough occured in September 2001 when Chudnovsky, Robertson, Seymour and Thomas announced that they could prove the Decomposition Conjecture in the following important special case.

THEOREM 2.17. (Chudnovsky, Robertson, Seymour and Thomas [**6**]) *If G is a Berge graph that contains the line graph of a bipartite subdivision of a 3-connected graph, then G has a balanced skew partition, or one of G or $\bar{G}$ has a 2-join or is the line graph of a bipartite graph.*

Eight months later, Chudnovsky, Robertson, Seymour and Thomas completed the proof of Theorem 2.15 and therefore of the Strong Perfect Graph Theorem. These results are contained in a monumental paper [**7**].

Conforti, Cornuéjols and Vušković [**18**] proved a weaker version of the Decomposition Conjecture where "skew partition" is replaced by "double star cutset". A *double star* is a node set S that contains two adjacent nodes u, v and a subset of the neighbors of u or v. Clearly, if G has a skew partition, then G has a double star cutset: Take $S = A \cup B$, $u \in A$ and $v \in B$. Although the decomposition result in [**18**] is weaker than Conjecture 2.14 for Berge graphs, it holds for a larger class of graphs than Berge graphs: By changing the decomposition from "skew partition" to "double star cutset", the result can be obtained for all odd-hole-free graphs instead of just Berge graphs.

THEOREM 2.18. (Conforti, Cornuéjols and Vušković [**18**]) *If G is an odd-hole-free graph, then G is a bipartite graph or the line graph of a bipartite graph or the complement of the line graph of a bipartite graph, or G has a double star cutset or a 2-join.*

Decomposition Theorem 2.18 was used by Cornuéjols, Liu and Vušković [**22**] to construct a polynomial time recognition algorithm for perfect graphs. Independently, Chudnovsky and Seymour [**9**] found an algorithm for perfect graph recognition which does not use decomposition. Both algorithms [**9**], [**22**] build on the companion paper [**5**] which performs a certain "cleaning" step in polynomial time. A key step in the cleaning procedure [**5**] is due to Chudnovsky and Seymour.

2.4. Proof of Theorem 1.11. It suffices to prove that (iii) implies (i). Assume that M is a 0,1 matrix with no column of 0's such that

(iii) max $\{wx : \; Mx \leq \mathbf{1}, \; x \geq 0\}$ has an integral optimal solution x for all $w \in \{0,1\}^n$.

We can assume w.l.o.g. that M is a clutter matrix. Let $G(M)$ be the graph having a node for each column of M and an edge between nodes j and k if M has a row i with $m_{ij} = m_{ik} = 1$.

Claim 1: M is the clique-node matrix of graph $G(M)$.

Proof: Suppose not. Then there exists a clique K of size at least 3 in $G(M)$ such that, for every row i, $K \not\subseteq N_i := \{j : m_{ij} = 1\}$. Choose K to be minimal with this property. Then, for each $j \in K$ there exists a distinct row i_j such that $K \cap N_{i_j} = K \setminus \{j\}$.

Let $w_j = 1$ for $j \in K$ and 0 otherwise. Consider the solution $x_j = \frac{1}{|K|-1}$ for $j \in K$ and 0 otherwise. It satisfies the constraints $Mx \leq \mathbf{1}$, $x \geq 0$ and it has objective value $wx = \frac{|K|}{|K|-1} > 1$, contradicting the assumption that max $\{wx : \; Mx \leq \mathbf{1}, \; x \geq 0\}$ has an integral optimal solution (an integral solution has objective value at most 1 since $x_j = 1$ for at most one node of K). This proves Claim 1.

Claim 2: $G(M)$ is a perfect graph.

Proof: In the 70's, this claim was proved using Lovász's characterization of minimally imperfect graphs [**41**] or Padberg's theorem [**47**]. Since we now have the Strong Perfect Graph Theorem at our disposal, let us use it. Thus, supposing that $G(M)$ is not perfect, $G(M)$ has a node set H that induces an odd hole or the complement of an odd hole. Let $w_j = 1$ for $j \in H$ and 0 otherwise.

If H induces an odd hole, consider the solution $x_j = \frac{1}{2}$ for $j \in H$ and 0 otherwise. It satisfies the constraints $Mx \leq \mathbf{1}$, $x \geq 0$ and it has objective value $wx = \frac{|H|}{2}$, contradicting the assumption that max $\{wx : Mx \leq \mathbf{1},\ x \geq 0\}$ has an integral optimal solution (an integral solution is the characteristic vector of a stable set and thus it has objective value at most $\frac{|H|-1}{2}$ since a stable set intersects H in at most $\frac{|H|-1}{2}$ nodes).

If H induces the complement of an odd hole, consider the solution $x_j = \frac{2}{|H|-1}$ for $j \in H$ and 0 otherwise. It satisfies the constraints $Mx \leq \mathbf{1}$, $x \geq 0$ and it has objective value $wx = \frac{2|H|}{|H|-1} > 2$, contradicting the assumption that max $\{wx : Mx \leq \mathbf{1},\ x \geq 0\}$ has an integral optimal solution (a stable set intersects H in at most 2 nodes). This proves Claim 2.

Since $G(M)$ is perfect, the complement graph $\bar{G}(M)$ is also perfect by the Perfect Graph Theorem. Let w be a nonnegative integral vector. If $w_i = 0$, delete node i from $\bar{G}(M)$ (and remove all the edges incident with node i). If $w_i \geq 2$, replicate node i of $\bar{G}(M)$ so that the resulting graph has w_i copies of node i (including node i itself), pairwise connected. By the Replication Lemma, the resulting graph $\bar{W}(M)$ is perfect. Thus the maximum cardinality of a clique in $\bar{W}(M)$ equals the chromatic number of $\bar{W}(M)$, i.e. a minimum cover of the nodes of $\bar{W}(M)$ by stable sets. In terms of the matrix M, this max-min equation is

$$\begin{array}{llcll} \max & wx & = & \min & \sum_i y_i \\ & Mx \leq \mathbf{1} & & & yM \geq w \\ & x \in \{0,1\}^n & & & y \in Z_+^m. \end{array}$$

Therefore the linear program max $\{wx : Mx \leq \mathbf{1},\ x \geq 0\}$ has an integral optimal dual solution y. In other words,

(i) the linear system $Mx \leq \mathbf{1}$, $x \geq 0$ is TDI. □

3. Ideal Clutters

3.1. Blockers. A *transversal* of a clutter $\mathcal{C}$ is a set of vertices that intersects all the edges. The *blocker* $b(\mathcal{C})$ of a clutter $\mathcal{C}$ is the clutter with $V(\mathcal{C})$ as vertex set and the minimal transversals of $\mathcal{C}$ as edge set. That is, $E\big(b(\mathcal{C})\big)$ consists of the minimal members of $\{B \subseteq V(\mathcal{C}) : |B \cap A| \geq 1 \text{ for all } A \in E(\mathcal{C})\}$. In other words, the rows of $M\big(b(\mathcal{C})\big)$ are the minimal 0,1 vectors x^T such that x belongs to the polyhedron $P(\mathcal{C}) = \{x \geq 0 : M(\mathcal{C})x \geq \mathbf{1}\}$.

EXAMPLE 3.1. Let G be a graph and s, t be distinct nodes of G. If $\mathcal{C}$ is the clutter of st-paths, then $b(\mathcal{C})$ is the clutter of minimal st-cuts.

Edmonds and Fulkerson [**26**] observed that $b(b(\mathcal{C})) = \mathcal{C}$. Before proving this property, we make the following remark.

REMARK 3.2. Let $\mathcal{H}$ and $\mathcal{K}$ be two clutters defined on the same vertex set. If
(i) every edge of $\mathcal{H}$ contains an edge of $\mathcal{K}$ and
(ii) every edge of $\mathcal{K}$ contains an edge of $\mathcal{H}$,
then $\mathcal{H} = \mathcal{K}$.

THEOREM 3.3. *If $\mathcal{C}$ is a clutter, then $b(b(\mathcal{C})) = \mathcal{C}$.*

PROOF. Let A be an edge of $\mathcal{C}$. The definition of $b(\mathcal{C})$ implies that $|A \cap B| \geq 1$, for every edge B of $b(\mathcal{C})$. So A is a transversal of $b(\mathcal{C})$, i.e. A contains an edge of $b(b(\mathcal{C}))$.

Now let A be an edge of $b(b(\mathcal{C}))$. We claim that A contains an edge of $\mathcal{C}$. Suppose otherwise. Then $V(\mathcal{C}) - A$ is a transversal of $\mathcal{C}$ and therefore it contains an edge B of $b(\mathcal{C})$. But then $A \cap B = \emptyset$ contradicts the fact that A is an edge of $b(b(\mathcal{C}))$. So the claim holds.

Now the theorem follows from Remark 3.2. □

Two 0,1 matrices of the form $M(\mathcal{C})$ and $M(b(\mathcal{C}))$ are said to form a *blocking pair*. The next theorem is an important result due to Lehman [**38**]. It states that, for a blocking pair A, B of 0,1 matrices, the polyhedron P defined by

$$Ax \geq \mathbf{1} \tag{3.1}$$

$$x \geq 0 \tag{3.2}$$

is integral if and only if the polyhedron Q defined by

$$Bx \geq \mathbf{1} \tag{3.3}$$

$$x \geq 0 \tag{3.4}$$

is integral. The proof of this result uses the following remark.

REMARK 3.4.
(i) The rows of B are exactly the 0,1 extreme points of P.
(ii) If an extreme point x of P satisfies $x^T \geq \lambda^T B$ where $\lambda_i \geq 0$ and $\sum \lambda_i = 1$, then x is a 0,1 extreme point of P.

PROOF. (i) follows from the fact that the rows of B are the minimal 0,1 vectors in P.

To prove (ii), note that x is an extreme point of $P_I = \{\chi : \chi^T \geq \lambda^T B$ where $\lambda_i \geq 0$ and $\sum \lambda_i = 1\}$ for otherwise x would be a convex combination of distinct $x^1, x^2 \in P_I$ and, since $P_I \subseteq P$, this would contradict the assumption that x is an extreme point of P. Now (ii) follows by observing that the extreme points of P_I are exactly the rows of B. □

THEOREM 3.5. (Lehman [**38**]) *A clutter is ideal if and only if its blocker is.*

PROOF. By Theorem 3.3, it suffices to show that if P defined by (3.1)-(3.2) is integral, then Q defined by (3.3)-(3.4) is also integral.

Let a be an arbitrary extreme point of Q. By (3.3), $Ba \geq \mathbf{1}$, i.e. $a^T x \geq 1$ is satisfied by every x such that x^T is a row of B. Since P is an integral polyhedron, it follows from Remark 3.4(i) that $a^T x \geq 1$ is satisfied by all the extreme points of P. By (3.4), $a \geq 0$. Therefore $a^T x \geq 1$ is satisfied by all points in P. Furthermore, $a^T x = 1$ for some $x \in P$. Now, by linear programming duality, we have

$$1 \quad = \quad \min\{a^T x : x \in P\} \quad = \quad \max\{\lambda^T \mathbf{1} : \lambda^T A \leq a^T,\ \lambda \geq 0\}.$$

Therefore, by Remark 3.4(ii), a is a 0,1 extreme point of Q. □

3.2. Deletion, Contraction and Minors. Let $\mathcal{C}$ be a clutter. For $j \in V(\mathcal{C})$, the *contraction* $\mathcal{C}/j$ and *deletion* $\mathcal{C} \setminus j$ are clutters defined as follows: Both have $V(\mathcal{C}) - \{j\}$ as vertex set, $E(\mathcal{C}/j)$ is the set of minimal members in $\{S - \{j\} : S \in E(\mathcal{C})\}$ and $E(\mathcal{C} \setminus j) = \{S : j \notin S \in E(\mathcal{C})\}$.

EXAMPLE 3.6. Let $\mathcal{C}$ be the clutter of st-paths in a graph G where s, t are distinct nodes of G, and let j be an edge of G. Then $\mathcal{C}/j$ is the clutter of st-paths in the graph obtained from G by contracting edge j in the graph theoretical sense (West [**56**] page 65) and $\mathcal{C} \setminus j$ is the clutter of st-paths in the graph obtained from G by deleting the edge j in the graphical sense. See Figure 5 for an example.

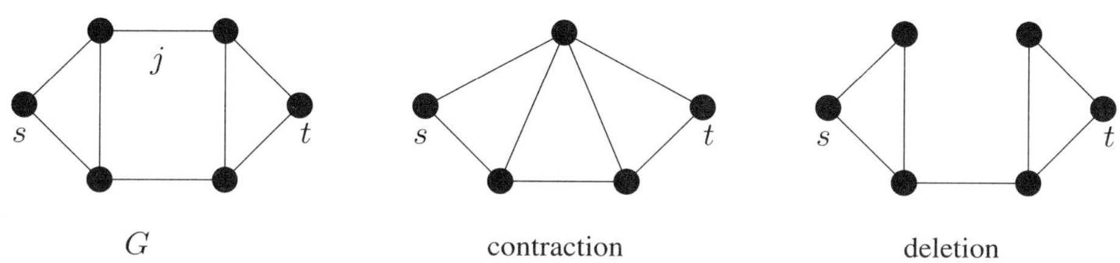

FIGURE 5. Examples of contraction and deletion in a graph.

Let $\mathcal{C}$ be a clutter and let M be the corresponding clutter matrix. Contracting $j \in V(\mathcal{C})$ corresponds to removing column j from matrix M as well as the resulting dominating rows. In the set covering constraints $Mx \geq \mathbf{1}$ of (1.1), this corresponds to setting $x_j = 0$. Deleting j corresponds to removing column j from M as well as all the rows with a 1 in column j. This corresponds to setting $x_j = 1$ in $Mx \geq \mathbf{1}$.

EXAMPLE 3.7. For the graphs of Figure 5, the clutters of st-paths have the following clutter matrices, where j corresponds to the last column of matrix M.

$$M = \begin{pmatrix} 1 & 0 & 0 & 0 & 0 & 1 & 0 & 1 \\ 1 & 0 & 0 & 0 & 1 & 0 & 1 & 1 \\ 0 & 1 & 1 & 0 & 0 & 1 & 0 & 1 \\ 0 & 1 & 1 & 0 & 1 & 0 & 1 & 1 \\ 0 & 1 & 0 & 1 & 0 & 0 & 1 & 0 \\ 0 & 1 & 0 & 1 & 1 & 1 & 0 & 0 \\ 1 & 0 & 1 & 1 & 0 & 0 & 1 & 0 \\ 1 & 0 & 1 & 1 & 1 & 1 & 0 & 0 \end{pmatrix},$$

$$M/j = \begin{pmatrix} 1 & 0 & 0 & 0 & 0 & 1 & 0 \\ 1 & 0 & 0 & 0 & 1 & 0 & 1 \\ 0 & 1 & 1 & 0 & 0 & 1 & 0 \\ 0 & 1 & 1 & 0 & 1 & 0 & 1 \\ 0 & 1 & 0 & 1 & 0 & 0 & 1 \\ 0 & 1 & 0 & 1 & 1 & 1 & 0 \\ 1 & 0 & 1 & 1 & 0 & 0 & 1 \end{pmatrix}, \quad M \setminus j = \begin{pmatrix} 0 & 1 & 0 & 1 & 0 & 0 & 1 \\ 0 & 1 & 0 & 1 & 1 & 1 & 0 \\ 1 & 0 & 1 & 1 & 0 & 0 & 1 \\ 1 & 0 & 1 & 1 & 1 & 1 & 0 \end{pmatrix}.$$

Contractions and deletions of distinct vertices can be performed sequentially, and it is easy to show that the result does not depend on the order.

PROPOSITION 3.8. *For a clutter $\mathcal{C}$ and distinct vertices j_1, j_2,*

(i) $(\mathcal{C} \setminus j_1) \setminus j_2 = (\mathcal{C} \setminus j_2) \setminus j_1$

(ii) $(\mathcal{C}/j_1)/j_2 = (\mathcal{C}/j_2)/j_1$
(iii) $(\mathcal{C}\backslash j_1)/j_2 = (\mathcal{C}/j_2)\backslash j_1$

PROOF. Use the definitions of contraction and deletion. □

DEFINITION 3.9. A clutter $\mathcal{D}$ obtained from $\mathcal{C}$ by a sequence of deletions and contractions is a *minor* of $\mathcal{C}$.

If V_1 and V_2 are disjoint subsets of $V(\mathcal{C})$, we let $\mathcal{C}/V_1\backslash V_2$ be the minor obtained from $\mathcal{C}$ by contracting all vertices of V_1 and deleting all vertices of V_2. If $V_1 \neq \emptyset$ or $V_2 \neq \emptyset$, the minor is *proper*.

PROPOSITION 3.10. *For a clutter $\mathcal{C}$ and $U \subset V(\mathcal{C})$,*

(i) $b(\mathcal{C}\backslash U) = b(\mathcal{C})/U$
(ii) $b(\mathcal{C}/U) = b(\mathcal{C})\backslash U$

PROOF. Use the definitions of contraction, deletion and blocker. □

We leave it as an exercise to prove the following result.

PROPOSITION 3.11. *If a clutter is ideal, then so are all its minors.*

COROLLARY 3.12. *Let M be a 0,1 matrix. The following are equivalent.*

(i) *The polyhedron $\{x \geq \mathbf{0},\ Mx \geq \mathbf{1}\}$ is integral.*
(ii) *The bounded polyhedron $\{0 \leq x \leq \mathbf{1},\ Mx \geq \mathbf{1}\}$ is integral.*

3.3. *st*-cuts and *st*-paths. Consider a graph G and distinct nodes $s, t \in V(G)$. Let $\mathcal{C}$ be the clutter where $V(\mathcal{C}) = E(G)$ and $E(\mathcal{C})$ is the family of st-paths of G.

THEOREM 3.13. (Ford and Fulkerson [**30**]) *The clutter $\mathcal{C}$ of st-paths has the MFMC property.*

This theorem is a restatement of the famous Max Flow Min Cut theorem of Ford-Fulkerson: For any vector $w \in \mathbb{Z}_+^{E(G)}$ of capacities, the minimum capacity of an st-cut equals the maximum number of st-paths such that every edge $e \in E(G)$ belongs to at most w_e of the paths. Indeed, the Ford-Fulkerson theorem states that both (1.1) and (1.2) have optimal solutions that are integral whenever the capacity vector w is integral.

Theorem 3.13 implies that $\mathcal{C}$ is ideal and therefore

$$\{x \in \mathbb{R}_+^{E(G)} : x(P) \geq 1 \text{ for all } st\text{-paths } P\}$$

is an integral polyhedron. Its extreme points are the minimal st-cuts. It will be convenient to refer to minimal st-cuts simply as st-cuts.

As a consequence of Lehman's theorem (Theorem 3.5), the clutter of st-cuts is also ideal. So

$$\{x \in \mathbb{R}_+^{E(G)} : x(C) \geq 1 \text{ for all } st\text{-cuts } C\}$$

is an integral polyhedron. In fact, it is easy to show that the clutter of st-cuts has the MFMC property.

3.4. T-cuts and T-joins. Consider a connected graph G with nonnegative edge weights w_e, for $e \in E(G)$. The Chinese Postman Problem consists in finding a minimum weight closed walk going through each edge at least once (the edges of the graph represent roads where mail must be delivered and w_e is the length of the road). Equivalently, the postman must find a minimum weight set of edges $J \subseteq E(G)$ such that $J \cup E(G)$ induces an Eulerian graph, i.e. J induces a graph whose odd degree nodes coincide with the odd degree nodes of G. Since $w \geq 0$, we can assume w.l.o.g. that J is acyclic. Such an edge set J is called a *postman set.*

The problem is generalized as follows. Let G be a graph and T a node set of G of even cardinality. An edge set J of G is called a *T-join* if it induces an acyclic graph the odd degree nodes of which coincide with T. For disjoint node sets S_1, S_2, let (S_1, S_2) denote the set of edges with one endnode in S_1 and the other in S_2. A *T-cut* is a minimal edge set of the form $(S, V(G) - S)$ where S is a set of nodes with $|T \cap S|$ odd. Clearly every T-cut intersects every T-join. Note that when $|T| = 2$, say $T = \{s,t\}$, T-joins and T-cuts are nothing but st-paths and st-cuts. Therefore the theory of T-joins and T-cuts is a generalization of Ford and Fulkerson's network flow theory. In this section, we will show that the clutters of T-joins and T-cuts are ideal.

Edmonds and Johnson [**28**] considered the problem of finding a minimum weight T-join. One way to solve this problem is to reduce it to the perfect matching problem in a complete graph K_p, where $p = |T|$. Namely, compute the lengths of shortest paths in G between all pairs of nodes in T, use these values as edge weights in K_p and find a minimum weight perfect matching in K_p. The union of the corresponding paths in G is a minimum weight T-join. Edmonds and Johnson developed a direct primal-dual algorithm for the minimum weight T-join problem and, as a by-product, obtained that the clutter of T-cuts is ideal.

THEOREM 3.14. (Edmonds and Johnson [**28**]) *The polyhedron*

$$x(C) \geq 1 \quad \text{for all } T\text{-cuts } C \tag{3.5}$$
$$x_e \geq 0 \quad \text{for all } e \in E(G). \tag{3.6}$$

is an integral polyhedron.

In the next section, we give a non-algorithmic proof of this theorem suggested by Pulleyblank [**49**].

The Edmonds-Johnson theorem together with the fact that the blocker of an ideal clutter is ideal (Theorem 3.3 of Lehman) implies that the clutter of T-joins is also ideal. That is

$$x(J) \geq 1 \quad \text{for all } T\text{-joins } J$$
$$x_e \geq 0 \quad \text{for all } e \in E(G).$$

is an integral polyhedron.

The clutter of T-cuts does not pack, nor does the clutter of T-joins. The reader is encouraged to find examples showing this.

Proof of the Edmonds-Johnson Theorem

First, we prove the following lemma. For $v \in V(G)$, let $\delta(v)$ denote the set of edges incident with v. A *star* is a tree where one node is adjacent to all the other nodes.

LEMMA 3.15. *Let $\tilde{x}$ be an extreme point of the polyhedron*

$$x(\delta(v)) \geq 1 \quad \textit{for all} \quad v \in T \tag{3.7}$$

$$x_e \geq 0 \quad \textit{for all} \quad e \in E(G). \tag{3.8}$$

The connected components of the graph $\tilde{G}$ induced by the edges such that $\tilde{x}_e > 0$ are either

(i) *odd cycles with nodes in T and edges $\tilde{x}_e = 1/2$, or*
(ii) *stars with nodes in T, except possibly the center, and edges $\tilde{x}_e = 1$.*

PROOF. Every connected component C of $\tilde{G}$ is either a tree or contains a unique cycle, since the number of edges in C is at most the number of inequalities (3.7) that hold with equality.

Assume first that C contains a unique cycle. Then (3.7) holds with equality for all nodes of C, which are therefore in T. Now C is a cycle since, otherwise, C has a pendant edge e with $\tilde{x}_e = 1$ and therefore C is disconnected, a contradiction. If C is an even cycle, then by alternately increasing and decreasing $\tilde{x}$ around the cycle by a small ϵ ($-\epsilon$ respectively), $\tilde{x}$ can be written as a convex combination of two points satisfying (3.7) and (3.8). So (i) must hold.

Assume now that C is a tree. Then (3.7) holds with equality for at least $|V(C)| - 1$ nodes of C. In particular, it holds with equality for at least one node of degree one. Since C is connected, this implies that C is a star and (ii) holds. □

Proof of Theorem 3.14: In order to prove the theorem, it suffices to show that every extreme point $\tilde{x}$ of the polyhedron (3.5)–(3.6) is the characteristic vector of a T-join. We proceed by induction on the number of nodes of G.

Suppose first that $\tilde{x}$ is an extreme point of the polyhedron (3.7)–(3.8). Consider a connected component of the graph $\tilde{G}$ induced by the edges such that $\tilde{x}_e > 0$ and let S be its node set. Since $\tilde{x}(S, V(G) - S) = 0$, it follows from (3.5) that S contains an even number of nodes of T. By Lemma 3.15, $\tilde{G}$ contains no odd cycle, showing that $\tilde{x}$ is an integral vector. Furthermore, $\tilde{x}$ is the characteristic vector of a T-join since, by Lemma 3.15 again, the component of $\tilde{G}$ induced by S is a star and $|S \cap T|$ even implies that the center is in T if and only if the star has an odd number of edges.

Assume now that $\tilde{x}$ is not an extreme point of the polyhedron (3.7)–(3.8). Then there is some T-cut $C = (V_1, V_2)$ with $|V_1| \geq 2$ and $|V_2| \geq 2$ such that

$$\tilde{x}(C) = 1.$$

Let $G_1 = (V_1 \cup \{v_2\}, E_1)$ be the graph obtained from G by contracting V_2 to a single node v_2. Similarly, $G_2 = (V_2 \cup \{v_1\}, E_2)$ is the graph obtained from G by contracting V_1 to a single node v_1. The new nodes v_1, v_2 belong to T. For $i = 1, 2$, let $\tilde{x}^i$ be the restriction of $\tilde{x}$ to E_i. Since every T-cut of G_i is also a T-cut of G, it follows by induction that $\tilde{x}^i$ is greater than or equal to a convex combination of characteristic vectors of T-joins of G_i. Let $\mathcal{T}_i$ be this set of T-joins. Each T-join in $\mathcal{T}_i$ has exactly one edge incident with v_i. Since $\tilde{x}^1$ and $\tilde{x}^2$ coincide on the edges of C, it follows that the T-joins of $\mathcal{T}_1$ can be combined with those of $\mathcal{T}_2$ to form T-joins of G and that $\tilde{x}$ is greater than or equal to a convex combination of characteristic vectors of T-joins of G. Since $\tilde{x}$ is an extreme point, it is the characteristic vector of a T-join. □

We have just proved that the clutter of T-cuts is ideal. It does not have the MFMC property in general graphs. However Seymour proved that it does in bipartite graphs. Seymour also showed that, in a general graph, if the edge weights w_e are integral and their sum is even in every cycle, then the dual variables y in (1.2) can be chosen to be integral in an optimum solution.

3.5. Odd Cycles in Graphs. In this section, we consider the clutter $\mathcal{O}_G$ of odd cycles in a graph G, namely $V(\mathcal{O}_G) = E(G)$ and $E(\mathcal{O}_G)$ is the family of edge sets of the odd cycles in G.

EXAMPLE 3.16. For the complete graph K_4 on four nodes, the clutter of odd cycles has six vertices (the six edges of K_4) and four edges (the four triangles of K_4). The reader can check that its clutter matrix is Q_6. Recall that Q_6 is ideal but does not have the MFMC property.

For the complete graph K_5 on five nodes, the clutter of odd cycles has ten vertices (the ten edges of K_5) and its edges comprise all the triangles (there are ten) and all the pentagons (there are twelve). Note that $x_j = \frac{1}{3}$ for $j = 1, \ldots, 10$ is an extreme point of the polyhedron $\{x \in \mathbb{R}^{10}_+ : \mathcal{O}_{K_5} x \geq \mathbf{1}\}$ and it is obviously nonintegral. Therefore $\mathcal{O}_{K_5}$ is not ideal.

Seymour [**54**] characterized exactly the graphs G for which $\mathcal{O}_G$ has the MFMC property and Guenin [**35**] characterized exactly when $\mathcal{O}_G$ is ideal.

THEOREM 3.17. (Seymour [**54**]) *The clutter of odd cycles in a graph G has the MFMC property if and only if it contains no Q_6 minor.*

THEOREM 3.18. (Guenin [**35**]) *The clutter of odd cycles in a graph G is ideal if and only if it contains no $\mathcal{O}_{K_5}$ minor.*

A proof of these theorems can be found in Chapter 75 of Schrijver's book on *Combinatorial Optimization* [**53**].

3.6. Minimally Nonideal Matrices. Lehman (Theorem 3.5) showed that ideal 0,1 matrices always come in pairs (if M is ideal, so is its blocker $b(M)$). Another important result of Lehman about ideal 0,1 matrices is the following theorem, which is reminiscent of Theorem 1.11 on perfect matrices.

THEOREM 3.19. (Lehman [**39**]) *For a 0,1 matrix A, the following statements are equivalent:*

(i) *the matrix A is ideal,*
(ii) *min $\{cx : \ Ax \geq \mathbf{1}, \ x \geq \mathbf{0}\}$ has an integral optimal solution x for all $c \in \{0, 1, +\infty\}^n$.*

The fact that (i) implies (ii) is an immediate consequence of the definition of idealness. The difficult part of Lehman's theorem is that (ii) implies (i). The main purpose of this section is to prove this result. This is done by studying properties of minimally nonideal matrices.

Lehman's Characterization

A 0,1 matrix A is *minimally nonideal* (*mni*) if

(i) A is a clutter matrix,
(ii) $Q(A) = \{x \geq 0 : \ Ax \geq \mathbf{1}\}$ is not an integral polyhedron,

(iii) For every $i = 1, \dots, n$, both $Q(A) \cap \{x : x_i = 0\}$ and $Q(A) \cap \{x : x_i = 1\}$ are integral polyhedra.

If A is *mni*, the clutter $\mathcal{C}(A)$ is also called *mni*. Equivalently, a clutter $\mathcal{C}$ is *mni* if it is not ideal but all its proper minors are ideal.

For $t \geq 2$ integer, let $\mathcal{J}_t$ denote the clutter with $t+1$ vertices and edges corresponding, respectively, to the points and lines of the finite degenerate projective plane. Namely, $V(\mathcal{J}_t) = \{0, \dots, t\}$, and $E(\mathcal{J}_t) = \{\{1, \dots, t\}, \{0,1\}, \{0,2\}, \dots, \{0,t\}\}$.

A matrix A is *isomorphic* to a matrix B if B can be obtained from A by a permutation of rows and a permutation of columns.

Let J denote a square matrix all of whose entries are 1's, and let I be the identity matrix. Given a *mni* matrix A, let $\bar{x}$ be a nonintegral extreme point of the polyhedron $Q(A) = \{x \geq 0 : Ax \geq \mathbf{1}\}$, that is, at least one component $\bar{x}_j$ of $\bar{x}$ is fractional. The maximum row submatrix $\bar{A}$ of A such that $\bar{A}\bar{x} = \mathbf{1}$ is called a *core* of A. So A has a core for each nonintegral extreme point of $Q(A)$.

THEOREM 3.20. (Lehman [**39**]) *Let A be a mni matrix and $B = b(A)$. Then*

(i) *A has a unique core $\bar{A}$ and B has a unique core $\bar{B}$;*

(ii) *$\bar{A}$ and $\bar{B}$ are square matrices;*

(iii) *Either A is isomorphic to $M(\mathcal{J}_t)$, $t \geq 2$, or the rows of $\bar{A}$ and $\bar{B}$ can be permuted so that*

$$\bar{A}\bar{B}^T = J + dI$$

for some positive integer d.

Lehman's proof of this theorem is rather terse. Seymour [**55**], Padberg [**48**] and Gasparyan, Preissmann and Sebö [**34**] give more accessible presentations of Lehman's proof. In the next section, we present a proof of Lehman's theorem following Padberg's polyhedral point of view. Before proving Theorem 3.20, we present some of its consequences.

Bridges and Ryser [**3**] studied square matrices Y, Z that satisfy the matrix equation $YZ = J + dI$.

THEOREM 3.21. (Bridges and Ryser [**3**]) *Let Y and Z be $n \times n$ 0,1 matrices such that $YZ = J + dI$ for some positive integer d. Then*

(i) *each row and column of Y has the same number r of ones, each row and column of Z has the same number s of ones with $rs = n + d$,*

(ii) *$YZ = ZY$,*

PROOF. It is straightforward to check that $(J + dI)^{-1} = \frac{1}{d}I - \frac{1}{d(n+d)}J$. Hence

$$YZ = J + dI \Rightarrow YZ(\frac{1}{d}I - \frac{1}{d(n+d)}J) = I \Rightarrow Z(\frac{1}{d}I - \frac{1}{d(n+d)}J)Y = I$$

$$\text{i.e.} \qquad ZY = \frac{1}{n+d}ZJY + dI = \frac{1}{n+d}\mathbf{s}\mathbf{r}^T + dI$$

where $\mathbf{s} = Z\mathbf{1}$ and $\mathbf{r} = Y^T\mathbf{1}$.

It follows that, for each i and j, $n + d$ divides $r_i s_j$. On the other hand, the trace of the matrix ZY is equal to the trace of YZ, which is $n(d+1)$. This implies $\frac{1}{n+d}(\sum_1^n s_i r_i) = n$ and, since $s_i > 0$ and $r_i > 0$, we have $r_i s_i = n + d$. Now consider distinct i, j. Since $r_i s_i = r_j s_j = n + d$ and $n + d$ divides $r_i s_j$ and $r_j s_i$, it follows that $r_i = r_j$ and $s_i = s_j$. Therefore, all columns of Z have the same sum s and all rows of Y have the same sum r. Furthermore, $ZY = J + dI$ and, by symmetry, all columns of Y have the same sum and all rows of Z have the same sum. □

Theorems 3.20 and 3.21 have the following consequence.

COROLLARY 3.22. *Let A be a mni matrix nonisomorphic to $M(\mathcal{J}_t)$. Then it has a non-singular row submatrix $\bar{A}$ with exactly r ones in every row and column. Moreover, rows of A not in $\bar{A}$ have at least $r+1$ ones.*

This implies the next result, which is a restatement of Theorem 3.19.

COROLLARY 3.23. *Let A be a $0,1$ matrix. $Q(A) = \{x \in R^n_+ : Ax \geq \mathbf{1}\}$ is an integral polyhedron if and only if $min\{wx : x \in Q(A)\}$ has an integral optimal solution for all $w \in \{0, 1, \infty\}^n$.*

Note that Theorem 1.9 mentioned in the introduction follows from Corollary 3.23.

Proof of Lehman's Theorem

Let A be an $m \times n$ *mni* matrix, $\bar{x}$ a nonintegral extreme point of $Q(A) = \{x \in R^n_+ : Ax \geq \mathbf{1}\}$ and $\bar{A}$ a core of A. That is, $\bar{A}$ is the maximal row submatrix of A such that $\bar{A}\bar{x} = \mathbf{1}$. For simplicity of notation, assume that $\bar{A}$ corresponds to the first p rows of A, i.e. the entries of $\bar{A}$ are a_{ij} for $i = 1, \ldots, p$ and $j = 1, \ldots, n$. Since A is *mni*, every component of $\bar{x}$ is nonzero. Therefore $p \geq n$ and $\bar{A}$ has no row or column containing only 0's or only 1's.

The following easy result will be applied to the bipartite representation G of the 0,1 matrix $J - \bar{A}$ where J denotes the $p \times n$ matrix of all 1's, namely ij is an edge of G if and only if $a_{ij} = 0$, for $1 \leq i \leq p$ and $1 \leq j \leq n$. Let $d(u)$ denote the degree of node u.

LEMMA 3.24. (de Bruijn and Erdös [**24**]) *Let $(I \cup J, E)$ be a bipartite graph with no isolated node. If $|I| \geq |J|$ and $d(i) \geq d(j)$ for all $i \in I$, $j \in J$ such that $ij \in E$, then $|I| = |J|$ and $d(i) = d(j)$ for all $i \in I$, $j \in J$ such that $ij \in E$.*

PROOF. $|I| = \sum_{i \in I}(\sum_{j \in N(i)} \frac{1}{d(i)}) \leq \sum_{i \in I} \sum_{j \in N(i)} \frac{1}{d(j)} = \sum_{j \in J} \sum_{i \in N(j)} \frac{1}{d(j)} = |J|$. Now the hypothesis $|I| \geq |J|$ implies that equality holds throughout. So $|I| = |J|$ and $d(i) = d(j)$ for all $i \in I$, $j \in J$ such that $ij \in E$. □

The key to proving Lehman's theorem is the following lemma. We refer the reader to Sections 8.1 to 8.5 in Chapter 8 of Schrijver [**52**] for any undefined term related to polyhedra.

LEMMA 3.25. *$p = n$ and, if $a_{ij} = 0$ for $1 \leq i, j \leq n$, then row i and column j of $\bar{A}$ have the same number of ones.*

PROOF. Let x^j be defined by

$$x^j_k = \begin{cases} \bar{x}_k & \text{if} \quad k \neq j \\ 1 & \text{if} \quad k = j \end{cases}$$

and let F_j be the face of $Q(A) \cap \{x_j = 1\}$ of smallest dimension that contains x^j. Since A is *mni*, F_j is an integral polyhedron. The proof of the lemma will follow unexpectedly from computing the dimension of F_j.

The point x^j lies at the intersection of the hyperplanes in $\bar{A}x = \mathbf{1}$ such that $a_{kj} = 0$ (at least $n - \sum_{k=1}^{p} a_{kj}$ such hyperplanes are independent since $\bar{A}$ has rank

n) and of the hyperplane $x_j = 1$ (independent of the previous hyperplanes). It follows that

$$dim(F_j) \leq n - (n - \sum_{k=1}^{p} a_{kj} + 1) = \sum_{k=1}^{p} a_{kj} - 1$$

Choose a row a^i of $\bar{A}$ such that $a_{ij} = 0$. Since $x^j \in F_j$, it is greater than or equal to a convex combination of extreme points b^ℓ of F_j, say $x^j \geq \sum_{\ell=1}^{t} \gamma_\ell b^\ell$, where $\gamma > 0$ and $\sum \gamma_\ell = 1$.

$$1 = a^i x^j \geq \sum_{\ell=1}^{t} \gamma_\ell a^i b^\ell \geq 1 \tag{3.9}$$

Therefore, equality must hold throughout. In particular $a^i b^\ell = 1$ for $\ell = 1, \ldots, t$. Since b^ℓ is a 0,1 vector, it has exactly one nonzero entry in the set of columns k where $a_{ik} = 1$. Another consequence of the fact that equality holds in (3.9) is that $x_k^j = \sum_{\ell=1}^{t} \gamma_\ell b_k^\ell$ for every k where $a_{ik} = 1$. Now, since $x_k^j > 0$ for all k, it follows that F_j contains at least $\sum_{k=1}^{n} a_{ik}$ linearly independent points b^ℓ, i.e.

$$dim(F_j) \geq \sum_{k=1}^{n} a_{ik} - 1.$$

Therefore, $\sum_{k=1}^{n} a_{ik} \leq \sum_{k=1}^{p} a_{kj}$ for all i, j such that $a_{ij} = 0$.

Now Lemma 3.24 applied to the bipartite representation of $J - \bar{A}$ implies that $p = n$ and

$$\sum_{k=1}^{n} a_{ik} = \sum_{k=1}^{n} a_{kj} \text{ for all } i, j \text{ such that } a_{ij} = 0.$$

□

LEMMA 3.26. *$\bar{x}$ has exactly n adjacent extreme points in $Q(A)$, all with 0,1 coordinates.*

PROOF. By Lemma 3.25, exactly n inequalities of $A\bar{x} \geq \mathbf{1}$ hold with equality, namely $\bar{A}\bar{x} = \mathbf{1}$. In the polyhedron $Q(A)$, an edge adjacent to $\bar{x}$ is defined by $n-1$ of the n equalities in $\bar{A}x = \mathbf{1}$. Moving along such an edge from $\bar{x}$, at least one of the coordinates decreases. Since $Q(A) \in R_+^n$, this implies that $\bar{x}$ has exactly n adjacent extreme points on $Q(A)$. Suppose $\bar{x}$ has a nonintegral adjacent extreme point $\bar{x}'$. Since A is *mni*, $0 < \bar{x}'_j < 1$ for all j. Let $\bar{A}'$ be the $n \times n$ nonsingular submatrix of A such that $\bar{A}'\bar{x}' = \mathbf{1}$. Since $\bar{x}$ and $\bar{x}'$ are adjacent on $Q(A)$, $\bar{A}$ and $\bar{A}'$ differ in only one row. W.l.o.g. assume that $\bar{A}'$ corresponds to rows 2 to $n+1$. Since A contains no dominating row, there exists j such that $a_{1j} = 0$ and $a_{n+1,j} = 1$. Since $\bar{A}'$ cannot contain a column with only 1's, $a_{ij} = 0$ for some $2 \leq i \leq n$. But now, Lemma 3.24 is contradicted with row i and column j in either $\bar{A}$ or $\bar{A}'$. □

Lemma 3.26 has the following implication. Let $\bar{B}$ denote the $n \times n$ 0,1 matrix whose rows are the extreme points of $Q(A)$ adjacent to $\bar{x}$. By Remark 3.4(i), $\bar{B}$ is a submatrix of B. By Lemma 3.26, $\bar{B}$ satisfies the matrix equation

$$\bar{A}\bar{B}^T = J + D$$

where J is the matrix of all 1's and D is a diagonal matrix with positive diagonal entries $d_1, \ldots, d_n$.

LEMMA 3.27. *Either*

(i) $\bar{A} = \bar{B}$ *are isomorphic to* $M(\mathcal{J}_t)$, *for* $t \geq 2$, *or*
(ii) $D = dI$, *where* d *is a positive integer.*

PROOF. Consider the bipartite representation G of the 0,1 matrix $J - \bar{A}$.

Case 1: G is connected.
Then it follows from Lemma 3.25 that

$$\sum_k a_{ik} = \sum_k a_{kj} \text{ for all } i, j. \tag{3.10}$$

Let α denote this common row and column sum.

$$(n + d_1, \ldots, n + d_n) = \mathbf{1}^T(J + D) = \mathbf{1}^T \bar{A}\bar{B}^T = (\mathbf{1}^T \bar{A})\bar{B}^T = \alpha \mathbf{1}^T \bar{B}^T$$

Since there is at most one d, $1 \leq d < \alpha$, such that $n + d$ is a multiple of α, all d_i must be equal to d, i.e. $D = dI$.

Case 2: G is disconnected.
Let $q \geq 2$ denote the number of connected components in G and let

$$\bar{A} = \begin{pmatrix} K_1 & & \mathbf{1} \\ & \cdots & \\ \mathbf{1} & & K_q \end{pmatrix}$$

where K_t are 0,1 matrices, for $t = 1, \ldots, q$. It follows from Lemma 3.25 that the matrices K_t are square and $\sum_k a_{ik} = \sum_k a_{kj} = \alpha_t$ in each K_t.

Suppose first that $\bar{A}$ has no row with $n - 1$ ones. Then every K_t has at least two rows and columns. We claim that, for every j, k, there exist i, l such that $a_{ij} = a_{ik} = a_{lj} = a_{lk} = 1$. The claim is true if $q \geq 3$ or if $q = 2$ and j, k are in the same component (simply take two rows i, l from a different component). So suppose $q = 2$, column j is in K_1 and column k is in K_2. Since no two rows are identical, we must have $\alpha_1 \geq 1$, i.e. $a_{ij} = 1$ for some row i of K_1. Similarly, $a_{lk} = 1$ for some row l of K_2. The claim follows.

For each row b of $\bar{B}$, the vector $\bar{A}b^T$ has an entry greater than or equal to 2, so there exist two columns j, k such that $b_j = b_k = 1$. By the claim, there exist rows a_i and a_l of $\bar{A}$ such that $a_i b^T \geq 2$ and $a_l b^T \geq 2$, contradicting the fact that $\bar{A}b^T$ has exactly one entry greater than 1.

Therefore $\bar{A}$ has a row with $n - 1$ ones. Now it is routine to check that $\bar{A}$ is isomorphic to $M(\mathcal{J}_t)$, for $t \geq 2$. □

To complete the proof of Theorem 3.20, it only remains to show that the core $\bar{A}$ is unique and that $\bar{B}$ is a core of B and is unique.

If $\bar{A} = M(\mathcal{J}_t)$ for some $t \geq 2$, then the fact that A has no dominated rows implies that $A = \bar{A}$. Thus $B = \bar{B} = M(\mathcal{J}_t)$. So, the theorem holds in this case.

If $\bar{A}\bar{B}^T = J + dI$ for some positive integer d, then, by Theorem 3.21, all rows of $\bar{A}$ contain r ones. Therefore, $\bar{x}_j = \frac{1}{r}$, for $j = 1, \ldots, n$. The feasibility of $\bar{x}$ implies that all rows of A have at least r ones, and Lemma 3.25 implies that exactly n rows of A have r ones. Now $Q(A)$ cannot have a nonintegral extreme point $\bar{x}'$ distinct from $\bar{x}$, since the above argument applies to $\bar{x}'$ as well. Therefore A has a unique core $\bar{A}$. Since $\bar{x}$ has exactly n neighbors in $Q(A)$ and they all have s components equal to one, the inequality $\sum_1^n x_i \geq s$ is valid for the 0,1 points in $Q(A)$. This shows that every row of B has at least s ones and exactly n rows of B have s ones. Since B is *mni*, $\bar{B}$ is the unique core of B. □

Examples of *mni* Clutters

Let $Z_n = \{0, \ldots, n-1\}$. We define addition of elements in Z_n to be addition modulo n. Let $k \leq n-1$ be a positive integer. For each $i \in Z_n$, let C_i denote the subset $\{i, i+1, \ldots, i+k-1\}$ of Z_n. Define the *circulant* clutter $\mathcal{C}_n^k$ by $V(\mathcal{C}_n^k) = Z_n$ and $E(\mathcal{C}_n^k) = \{C_0, \ldots, C_{n-1}\}$.

Lehman [**38**] gave three infinite classes of minimally nonideal clutters: $\mathcal{C}_n^2$, $n \geq 3$ odd, their blockers, and the degenerate projective planes $\mathcal{J}_n$, $n \geq 2$.

CONJECTURE 3.28. (Cornuéjols and Novick [**23**]) *There exists n_0 such that, for $n \geq n_0$, all mni matrices have a core isomorphic to $\mathcal{C}_n^2$, $\mathcal{C}_n^{\frac{n+1}{2}}$ for $n \geq 3$ odd, or $\mathcal{J}_n$, for $n \geq 2$.*

However, there exist several known "small" *mni* matrices that do not belong to any of the above classes. For example, Lehman [**38**] noted that $\mathcal{F}_7$ is *mni*. $\mathcal{F}_7$ is the clutter with 7 vertices and 7 edges corresponding to points and lines of the Fano plane (finite projective geometry on 7 points):

$$M(\mathcal{F}_7) = \begin{pmatrix} 1 & 1 & 0 & 1 & 0 & 0 & 0 \\ 0 & 1 & 1 & 0 & 1 & 0 & 0 \\ 0 & 0 & 1 & 1 & 0 & 1 & 0 \\ 0 & 0 & 0 & 1 & 1 & 0 & 1 \\ 1 & 0 & 0 & 0 & 1 & 1 & 0 \\ 0 & 1 & 0 & 0 & 0 & 1 & 1 \\ 1 & 0 & 1 & 0 & 0 & 0 & 1 \end{pmatrix}$$

Let K_5 denote the complete graph on five nodes and let $\mathcal{O}_{K_5}$ denote the clutter whose vertices are the edges of K_5 and whose edges are the odd cycles of K_5 (the triangles and the pentagons). Seymour [**54**] noted that $\mathcal{O}_{K_5}$, $b(\mathcal{O}_{K_5})$, and $\mathcal{C}_9^2$ with the extra edge $\{3, 6, 9\}$ are *mni*.

Ding [**25**] found the following *mni* clutter: $V(\mathcal{D}_8) = \{1, \ldots, 8\}$ and

$$E(\mathcal{D}_8) = \{\{1,2,6\},\{2,3,5\},\{3,4,8\},\{4,5,7\},\{2,5,6\},\{1,6,7\},\{4,7,8\},\{1,3,8\}\}.$$

Cornuéjols and Novick [**23**] characterized the *mni* circulant clutters $\mathcal{C}_n^k$. They showed that the following ten clutters are the only *mni* $\mathcal{C}_n^k$ for $k \geq 3$:

$$\mathcal{C}_5^3,\ \mathcal{C}_8^3,\ \mathcal{C}_{11}^3,\ \mathcal{C}_{14}^3,\ \mathcal{C}_{17}^3,\ \mathcal{C}_7^4,\ \mathcal{C}_{11}^4,\ \mathcal{C}_9^5,\ \mathcal{C}_{11}^6,\ \mathcal{C}_{13}^7.$$

Independently, Qi [**50**] discovered $\mathcal{C}_9^5$ and $\mathcal{C}_{11}^6$ and Ding [**25**] discovered $\mathcal{C}_8^3$.

Let $\mathcal{T}_{K_5}$ denote the clutter whose vertices are the edges of K_5 and whose edges are the triangles of K_5 (interestingly, $M(\mathcal{T}_{K_5})$ is also the node-node adjacency matrix of the Petersen graph). It can be shown that $\mathcal{T}_{K_5}$, $core(b(\mathcal{T}_{K_5}))$ and their blockers are *mni*. Often, when a *mni* clutter $\mathcal{H}$ has the property that $core(\mathcal{H})$ and $core(b(\mathcal{H}))$ are also *mni*, many more *mni* clutters can be constructed from $\mathcal{H}$ and from $b(\mathcal{H})$, see [**23**]. For example, Cornuéjols and Novick [**23**] have constructed more than one thousand mni clutters from $\mathcal{T}_{K_5}$. More results can be found in [**46**].

Lütolf and Margot [**42**] designed a computer program that enumerates possible cores of minimally nonideal matrices. It first enumerates the square 0,1 matrices Y, Z that satisfy the matrix equation $YZ = J + dI$, and then checks that the covering polyhedron has a unique nonintegral extreme point. Lütolf and Margot

[**42**] enumerated all square *mni* matrices of dimension at most 12×12 and found 20 such matrices (previously, only 15 were known); they found 13 new square *mni* matrices of dimensions 14×14 and 17×17; and they found 38 new nonsquare *mni* matrices with 11, 14 and 17 columns with nonisomorphic cores. The overwhelming majority of these examples have $d = 1$: Only three cores with $d = 2$ are known (namely $\mathcal{F}_7, \mathcal{T}_{K_5}$ and the core of its blocker) and none with $d \geq 3$.

4. Conclusion

In these notes we considered the min-max equation (1.1)=(1.2) and the max-min equation (1.3)=(1.4). We focused on the cases where the polyhedra in (1.1) and (1.3) are integral. When this occurs, the matrix A is called ideal and perfect respectively. There are striking similarities between perfect and ideal matrices, such as Theorems 1.11 and 3.19. But there are also important differences. For instance, if A is perfect then (1.4) has an integral optimal solution for every integral w, whereas if A is ideal it is not true that (1.3) has an integral optimal solution for every integral w (e.g. $A = Q_6$ and $w = 1$). The Strong Perfect Graph Theorem of Chudnovsky, Robertson, Seymour and Thomas together with Chvátal's connection between perfect graphs and perfect matrices imply an explicit characterization of all minimally imperfect matrices whereas the situation for minimally nonideal matrices looks more complex, due to the existence of numerous individual examples (see Section 3.6). For perfect matrices, there is a beautiful polyhedral theorem (Theorem 1.11), an "excluded structure" theorem (the Strong Perfect Graph Theorem) and a decomposition theorem (Theorem 2.15). For ideal matrices, no decomposition theorem is known and there is no excluded structure theorem that provides an explicit list of minimally nonideal matrices, although Theorem 3.20 provides a partial answer. Finally, there is a nice polyhedral theorem for ideal matrices (Theorem 3.19). The counterpart of this polyhedral theorem for matrices with the MFMC property is still open (Conjecture 1.7).

References

[1] K. Appel and W. Haken, Every planar map is four colorable. Part I: Discharging, *Illinois Journal of Mathematics 21* (1977) 429-490.

[2] C. Berge, Färbung von Graphen deren sämtliche bzw. deren ungerade Kreise starr sind (Zusammenfassung), *Wissenschaftliche Zeitschrift, Martin Luther Universität Halle-Wittenberg, Mathematisch- Naturwissenschaftliche Reihe 10* (1961) 114-115.

[3] W.G. Bridges and H.J. Ryser, Combinatorial designs and related systems, *J. Algebra 13* (1969) 432-446.

[4] M. Burlet and J. Fonlupt, Polynomial algorithm to recognize a Meyniel graph, *Annals of Discrete Mathematics 21* (1984) 225-252.

[5] M. Chudnovsky, G. Cornuéjols, X. Liu, P. Seymour and K. Vušković, Cleaning for Bergeness, preprint (January 2003).

[6] M. Chudnovsky, N. Robertson, P. Seymour and R. Thomas, presentation at the Workshop on Graph Colouring and Decomposition, Princeton, September 2001.

[7] M. Chudnovsky, N. Robertson, P. Seymour and R. Thomas, The strong perfect graph theorem, preprint (June 2002, revised October 2002 and February 2003).

[8] M. Chudnovsky and P. Seymour, private communication (January 2002).

[9] M. Chudnovsky and P. Seymour, Recognizing Berge graphs, preprint (January 2003).

[10] V. Chvátal, On certain polytopes associated with graphs, *Journal of Combinatorial Theory B 18* (1975) 138-154.

[11] V. Chvátal, Star-cutsets and perfect graphs, *Journal of Combinatorial Theory B 39* (1985) 189-199.

[12] V. Chvátal, J. Fonlupt, L. Sun and A. Zemirline, Recognizing dart-free perfect graphs, *SIAM Journal on Computing 31* (2002) 1315-1338.

[13] V. Chvátal and N. Sbihi, Bull-free Berge graphs are perfect, *Graphs and Combinatorics 3* (1987) 127-139.

[14] V. Chvátal and N. Sbihi, Recognizing claw-free Berge graphs, *Journal of Combinatorial Theory B 44* (1988) 154-176.

[15] M. Conforti and G. Cornuéjols, Clutters that pack and the Max Flow Min Cut property: A conjecture, *The Fourth Bellairs Workshop on Combinatorial Optimization*, W.R. Pulleyblank and F.B. Shepherd eds. (1993).

[16] M. Conforti and G. Cornuéjols, Graphs without odd holes, parachutes or proper wheels: a generalization of Meyniel graphs and of line graphs of bipartite graphs, *Journal of Combinatorial Theory B 87* (2003) 300-330.

[17] M. Conforti, G. Cornuéjols and K. Vušković, Square-free perfect graphs, *Journal of Combinatorial Theory B 90* (2004) 257-307.

[18] M. Conforti, G. Cornuéjols and K. Vušković, Decomposition of odd-hole-free graphs by double star cutsets and 2-joins, to appear in the special issue of *Discrete Mathematics* dedicated to the Brazilian Symposium on Graphs, Algorithms and Combinatorics, Fortaleza, Brazil, March 2001.

[19] M. Conforti and E.L. Johnson, Two Min-Max theorems for graphs noncontractible to a four wheel, preprint, IBM Thomas J. Watson Research Center (1987).

[20] G. Cornuéjols, *Combinatorial Optimization: Packing and Covering, CBMS-NSF Regional Conference Series in Applied Mathematics 74*, SIAM (2001).

[21] G. Cornuéjols and W.H. Cunningham, Composition for perfect graphs, *Discrete Mathematics 55* (1985) 245-254.

[22] G. Cornuéjols, X. Liu, and K. Vušković, A polynomial algorithm for recognizing perfect graphs, preprint (January 2003). Extended abstract in *Proceedings of the 44th Symposium on the Foundations of Computer Science* (2003) 20-27.

[23] G. Cornuéjols and B. Novick, Ideal 0,1 matrices, *Journal of Combinatorial Theory B 60* (1994) 145-157.

[24] N.G. de Bruijn and P. Erdös, On a combinatorial problem, *Proc. Kon. Ned. Akad. v. Wetensch 51* 1277-1279.

[25] G. Ding, personal communication (1989).

[26] J. Edmonds and D.R. Fulkerson, Bottleneck extrema, *Journal of Combinatorial Theory B 8* (1970) 299-306.

[27] J. Edmonds and R. Giles, A Min-Max relation for submodular functions on graphs, *Annals of Discrete Mathematics 1* (1977) 185-204.

[28] J. Edmonds and E.L. Johnson, Matchings, Euler tours and the Chinese postman problem, *Mathematical Programming 5* (1973) 88-124.

[29] J. Fonlupt and A. Zemirline, A polynomial recognition algorithm for perfect K_4-$\{e\}$-free graphs, rapport technique RT-16, Artemis, IMAG, Grenoble, France (1987).

[30] L.R. Ford, Jr and D.R. Fulkerson, *Flows in Networks*, Princeton University Press, Princeton (1962).

[31] D. R. Fulkerson, On the perfect graph theorem, *Mathematical Programming*, T. C. Hu and S. M. Robinson eds., Academic Press (1973) 69-76.

[32] M.R. Garey and D.S. Johnson, *Computers and Intractability: A Guide to the Theory of NP-Completeness*, Freeman (1979).

[33] G.S. Gasparyan, Minimal imperfect graphs: A simple approach, *Combinatorica 16* (1996) 209-212.

[34] G.S. Gasparyan, M. Preissmann and A. Sebö, Imperfect and nonideal clutters: A common approach, *Combinatorica 23* (2003) 283-302.

[35] B. Guenin, A characterization of weakly bipartite graphs, *Journal of Combinatorial Theory B 83* (2001) 112-168.

[36] C. T. Hoàng, Some properties of minimal imperfect graphs, *Discrete Mathematics 160* (1996) 165-175.

[37] D. König, Graphs and Matrices, *Mat. Fiz. Lapok 38* (1931) 116-119 (in Hungarian).

[38] A. Lehman, On the width-length inequality, mimeographic notes (1965), published: *Mathematical Programming 17* (1979) 403-417.

[39] A. Lehman, On the width-length inequality and degenerate projective planes, unpublished manuscript (1981), published: *Polyhedral Combinatorics* (W. Cook and P.D. Seymour eds.), *DIMACS Series in Discrete Mathematics and Theoretical Computer Science 1*, American Mathematical Society, Providence, R.I. (1990) 101-105.
[40] L. Lovász, Normal hypergraphs and the perfect graph conjecture, *Discrete Mathematics 2* (1972) 253-267.
[41] L. Lovász, A Characterization of perfect graphs, *Journal of Combinatorial Theory B 13* (1972) 95-98.
[42] C. Lütolf and F. Margot, A catalog of minimally nonideal matrices, *Mathematical Methods of Operations Research 47* (1998) 221-241.
[43] F. Maffray and B. Reed, A description of claw-free perfect graphs, *Journal of Combinatorial Theory B 75* (1999) 134-156.
[44] K. Menger, Zur allgemeinen Kurventheorie, *Fundamenta Mathematicae 10* (1927) 96-115.
[45] G. L. Nemhauser and L. A. Wolsey, *Integer and Combinatorial Optimization*, Wiley (1988).
[46] B. Novick, Ideal 0,1 matrices, Ph.D. dissertation, Carnegie Mellon University (1990).
[47] M. Padberg, Perfect zero-one matrices, *Mathematical Programming 6* (1974) 180-196.
[48] M.W. Padberg, Lehman's forbidden minor characterization of ideal 0,1 matrices, *Discrete Mathematics 111* (1993) 409-420.
[49] W.R. Pulleyblank, personal communication (1992).
[50] L. Qi, On the set covering polytope and the MFMC-clutter, preprint, School of Mathematics, University of New South Wales (1989).
[51] F. Roussel and P. Rubio, About skew partitions in minimal imperfect graphs, *Journal of Combinatorial Theory B 83* (2001) 171-190.
[52] A. Schrijver, *Theory of Linear and Integer Programming*, Wiley, New York (1986).
[53] A. Schrijver, *Combinatorial Optimization: Polyhedra and efficiency*, Springer, Berlin (2003).
[54] P. Seymour, The matroids with the Max-Flow Min-Cut property, *Journal of Combinatorial Theory B 23* (1977) 189-222.
[55] P.D. Seymour, On Lehman's width-length characterization, in: *Polyhedral Combinatorics* (W. Cook and P.D. Seymour eds.), *DIMACS Series in Discrete Mathematics and Theoretical Computer Science 1*, American Mathematical Society, Providence, R.I. (1990) 107-117.
[56] D. B. West, *Introduction to Graph Theory*, Prentice Hall (1996).

LIF, Université de Marseille, Luminy, 13288 Marseille, France and GSIA, Carnegie Mellon University, Pittsburgh, PA 15213, USA
E-mail address: `cornuejols@lidil.univ-mrs.fr` or `gc0v@andrew.cmu.edu`

Proceedings of Symposia in Applied Mathematics
Volume **61**, 2004

Integer programming duality

Jean B. Lasserre

ABSTRACT. We consider the integer program $\max\{c'x \mid Ax = b, x \in \mathbb{N}^n\}$. A formal parallel between linear programming and continuous integration on one side, and discrete summation on the other side, shows that a natural duality for integer programs can be derived from the $\mathbb{Z}$-transform and Brion and Vergne' s counting formula. Along the same lines, we also provide a discrete Farkas lemma and show that the existence of a nonnegative integral solution $x \in \mathbb{N}^n$ to $Ax = b$ can be tested via a linear program. The linear program dual is then interpreted in the light of an abstract dual involving superadditive functions.

1. Introduction

We here consider the integer programming problem, or *integer program* (IP)

$$\mathbb{P}_d \quad \rightarrow \quad f_d(b,c) := \max_{x\in\mathbb{Z}^n} \{\, c'x \mid \quad Ax = b; \quad x \geq 0\} \tag{1.1}$$

(where $A \in \mathbb{Z}^{m\times n}, c \in \mathbb{R}^n, b \in \mathbb{Z}^m$), a fundamental problem in optimization with many applications of interest. Solving $\mathbb{P}_d$ still remains a formidable challenge for large (or even some small) values of n, m. Indeed, in the computational complexity terminology, $\mathbb{P}_d$ is NP-hard. However, if the dimension n is *fixed* then $\mathbb{P}_d$ is solvable in time polynomial in the input size. For standard sources the interested reader is referred to e.g. Nemhauser and Wolsey [**25**], Schrijver [**26**], and Wolsey [**29**].

$\mathbb{P}_d$ is the discrete analogue of the *linear program* (LP)

$$\mathbb{P} \quad \rightarrow \quad f(b,c) := \max_{x\in\mathbb{R}^n} \{\, c'x \mid \quad Ax = b; \quad x \geq 0\}, \tag{1.2}$$

where $A \in \mathbb{R}^{m\times n}, c \in \mathbb{R}^n, b \in \mathbb{R}^m$. In contrast with $\mathbb{P}_d$, $\mathbb{P}$ is considered *easy*, in the sense that it can be solved in time polynomial in the input size of the problem, in particular with *interior points methods*. Moreover, the linear program $\mathbb{P}$ has a *dual* linear program

$$\mathbb{P}^* \quad \rightarrow \quad \min_{\lambda\in\mathbb{R}^m} \{\, b'\lambda \mid \quad A'\lambda \geq c\}, \tag{1.3}$$

where $A \in \mathbb{R}^{m\times n}, c \in \mathbb{R}^n, b \in \mathbb{R}^m$. In particular, linear programming duality states that (i) $c'x \leq b'\lambda$ for any two feasible solutions $x \in \mathbb{R}^n$ and $\lambda \in \mathbb{R}^m$ of $\mathbb{P}$ and $\mathbb{P}^*$ respectively, and (ii) $c'x^* = b'\lambda^*$ whenever $x^* \in \mathbb{R}^n$ and $\lambda^* \in \mathbb{R}^m$ are optimal solutions of $\mathbb{P}$ and $\mathbb{P}^*$ respectively.

1991 *Mathematics Subject Classification.* Primary 90C10; Secondary 90C05, 05A15.

Key words and phrases. linear programming; duality; integer programming; generating functions.

The integer program $\mathbb{P}_d$ also has the following abstract dual problem

$$(1.4) \qquad \mathbb{P}_d^* \quad \rightarrow \quad \min_{f \in \Delta} \{ f(b) \mid \quad f(A_j) \geq c_j \quad j = 1, \ldots, n\},$$

where Δ is a space of functions $f : \mathbb{R}^m \rightarrow \mathbb{R}$ which are *superadditive* and satisfy $f(0) = 0$; see e.g. Jeroslow [**14**], Johnson [**15**], Wolsey [**28, 29**] (who consider the case $Ax \leq b$) and the references therein. As for linear programming, we also have the *weak duality* property, that is, $c'x \leq f(b)$ for any two feasible solutions $x \in \mathbb{N}^n, f \in \Delta$ of $\mathbb{P}_d$ and $\mathbb{P}_d^*$ respectively. Indeed, if $x \in \mathbb{N}^n$ satisfies $Ax = b$, and $f \in \Delta$ satisfies $f(A_j) \geq c_j$ for all $j = 1, \ldots, n$, then

$$(1.5) \qquad \sum_{j=1}^n c_j x_j \leq \sum_{j=1}^n f(A_j) x_j \leq \sum_{j=1}^n f(A_j x_j) \leq f(\sum_{j=1}^n A_j x_j) = f(b),$$

where the last two inequalities both follow from the superadditivity of f. Finally, *strong duality* also holds because the *value function* $b \mapsto f_d(b, c)$ in (1.1) is superadditive and a feasible solution for $\mathbb{P}_d^*$, hence an optimal solution of $\mathbb{P}_d^*$ (see e.g. Wolsey [**28**]).

However, in contrast with the linear program $\mathbb{P}^*$, the dual problem $\mathbb{P}_d^*$ is rather *abstract* and not easy to handle. In particular, the analogues for $\mathbb{P}_d^*$ of the dual variables $\lambda \in \mathbb{R}^m$ of $\mathbb{P}^*$, are now superadditive functions $f \in \Delta$.

We claim that a natural *duality* for integer programs can be derived from the $\mathbb{Z}$-transform (or generating function) associated with the *counting* version (defined below) of the integer program $\mathbb{P}_d$. Results for counting problems, notably by Barvinok [**3**], Barvinok and Pommersheim [**4**], Pukhlikov and Khovanskii [**13**], and in particular, Brion and Vergne' s counting formula [**6**], will prove especially useful.

For this purpose, we will take a little detour and consider the four related problems $\mathbb{P}, \mathbb{P}_d, \mathbb{I}$ and $\mathbb{I}_d$ displayed in the diagram below, in which the integer program $\mathbb{P}_d$ appears in the upper right corner.

Continuous Optimization		**Discrete Optimization**
$\mathbb{P}: \quad f(b,c) := \max\ c'x$ s.t. $\left[\begin{array}{lcl} Ax & = & b \\ x & \in & \mathbb{R}_+^n \end{array}\right.$	$\longleftrightarrow$	$\mathbb{P}_d: \quad f_d(b,c) := \max\ c'x$ s.t. $\left[\begin{array}{lcl} Ax & = & b \\ x & \in & \mathbb{N}^n \end{array}\right.$
$\updownarrow$		$\updownarrow$
Integration		**Summation**
$\mathbb{I}: \quad \widehat{f}(b,c) := \int_{\Omega(b)} e^{c'x} ds$ $\Omega(b) := \left[\begin{array}{lcl} Ax & = & b \\ x & \in & \mathbb{R}_+^n \end{array}\right.$	$\longleftrightarrow$	$\mathbb{I}_d: \quad \widehat{f}_d(b,c) := \sum_{x \in \Omega(b)} e^{c'x}$ $\Omega(b) := \left[\begin{array}{lcl} Ax & = & b \\ x & \in & \mathbb{N}^n. \end{array}\right.$

Problem $\mathbb{I}$ (in which ds denotes the Lebesgue measure on the affine subspace $\{x \in \mathbb{R}^n \mid Ax = b\}$ that contains the convex polyhedron $\Omega(b)$) is the *integration* version of the linear program $\mathbb{P}$, whereas Problem $\mathbb{I}_d$ is the *counting* version of the (discrete) integer program $\mathbb{P}_d$.

Why do these four problems should help in analyzing $\mathbb{P}_d$? Because firstly, $\mathbb{P}$ and $\mathbb{I}$, as well as $\mathbb{P}_d$ and $\mathbb{I}_d$, are simply related, and in the same manner. Next, as we will see, the nice and complete duality results available for $\mathbb{P}, \mathbb{I}$ and $\mathbb{I}_d$, extend in a natural way to $\mathbb{P}_d$.

In fact, $\mathbb{I}$ and $\mathbb{I}_d$ are the respective formal analogues in the algebra $(+, \times)$, of $\mathbb{P}$ and $\mathbb{P}_d$ in the algebra $(\oplus, +)$, where in the latter, the addition $a \oplus b$ stands for $\max(a, b)$; indeed, the "max" in $\mathbb{P}$ and $\mathbb{P}_d$ can be seen as an *idempotent* integral (or, *Maslov integral*) in this algebra (see e.g. Litvinov et al. [**23**]). For a nice parallel between results in probability ($(+, \times)$ algebra) and optimization ($(\max, +)$ algebra), the reader is referred to Bacelli et al. [**2**, §9].

Moreover, $\mathbb{P}$ and $\mathbb{I}$, as well as $\mathbb{P}_d$ and $\mathbb{I}_d$, are simply related via

$$(1.6) \qquad f(b, c) = \lim_{r\to\infty} \frac{1}{r} \ln \widehat{f}(b, rc); \quad f_d(b, c) = \lim_{r\to\infty} \frac{1}{r} \ln \widehat{f}_d(b, rc),$$

a relationship that will be useful later.

Next, concerning *duality*, the standard *Legendre-Fenchel* transform which yields the usual LP dual $\mathbb{P}^*$ of $\mathbb{P}$ in (1.3), has a natural analogue for integration, the *Laplace transform*, and thus, the *inverse Laplace transform* problem (that we call $\mathbb{I}^*$) is the formal analogue of $\mathbb{P}^*$, and provides a nice duality for integration (although not usually presented in these terms). Finally, the $\mathbb{Z}$-*transform* is the obvious analogue for *summation* of the Laplace transform for integration. We will see that in the light of recent results in *counting problems*, it permits to establish a nice duality for $\mathbb{I}_d$ of the same vein as the duality for (continuous) integration, and by (1.6), it also provides a powerful tool to analyze the integer program $\mathbb{P}_d$.

Summary of content.

(a) We first review the duality principles that are available for $\mathbb{P}$, $\mathbb{I}$ and $\mathbb{I}_d$ and underline the parallels and connections between them. In particular, a fundamental difference between the continuous and discrete cases is that, in the former, the data appear as *coefficients* of the dual variables whereas in the latter, the *same* data appear as *exponents* of the dual variables. Consequently, the (discrete) $\mathbb{Z}$-transform has many more *poles* than the Laplace transform. While the Laplace transform has only *real* poles, the $\mathbb{Z}$-transform has additional *complex* poles associated with each real pole, which induces some *periodic* behavior, a well known phenomenon in number theory where the $\mathbb{Z}$-transform (or *generating function*) is a standard tool (see e.g. Iosevich [**12**], Mitrinovíc et al [**24**]). So, the procedure of inverting the Laplace transform or the $\mathbb{Z}$-transform (i.e. solving the dual problems $\mathbb{I}^*$ and $\mathbb{I}_d^*$) is basically of the same nature, a complex integral. However, it is significantly more complicated in the discrete case, due to the presence of these additional complex poles.

(b) Then we use results from (a) to analyze the discrete optimization problem $\mathbb{P}_d$. Central in the analysis is Brion and Vergne' s inverse formula [**6**] for counting problems. In particular, we provide a closed form expression of the optimal value $f_d(b, c)$ which highlights the special role played by the so-called *reduced-costs* of the linear program $\mathbb{P}$ and the *complex poles* of the $\mathbb{Z}$-transform associated with each basis of the linear program $\mathbb{P}$. We also show that each basis B of the linear program $\mathbb{P}$ provides exactly $\det(B)$ complex *dual* vectors in $\mathbb{C}^m$, the complex (periodic) analogues for $\mathbb{P}_d$ of the unique dual vector in $\mathbb{R}^m$ for $\mathbb{P}$, associated with the basis B. As in linear programming (but in a more complicated way), the optimal value $f_d(b, c)$ of $\mathbb{P}_d$ can be found by inspection of (certain sums of) reduced costs associated with each vertex of $\Omega(b)$.

(c) We finally provide a *discrete* Farkas Lemma for the existence of nonnegative integral solutions $x \in \mathbb{N}^n$ to $Ax = b$. Its form also confirms the special role of the $\mathbb{Z}$-transform described earlier. Moreover, it permits to check the existence of a nonnegative integral solution by solving a related linear program. The dual of this linear program is then interpreted in the light of the dual $\mathbb{P}_d^*$ in (1.4). In particular, from its dual variables, one may recover a supperadditive function f^*, optimal solution of $\mathbb{P}_d^*$.

2. Duality for the continuous problems $\mathbb{P}$ and $\mathbb{I}$

With $A \in \mathbb{R}^{m\times n}, b \in \mathbb{R}^m, c \in \mathbb{R}^n$, let $\Omega(b) \subset \mathbb{R}^n$ be the convex polyhedron

$$\Omega(b) := \{x \in \mathbb{R}^n \,|\, Ax = b; \quad x \geq 0\}, \tag{2.1}$$

and consider the standard linear program (LP) $\mathbb{P}$ defined in (1.2), and its associated *integration* version

$$\mathbb{I}: \quad \widehat{f}(b, c) := \int_{\Omega(b)} \mathrm{e}^{c'x}\, ds \tag{2.2}$$

where ds is the Lebesgue measure on the affine subspace $\{x \in \mathbb{R}^n \,|\, Ax = b\}$ that contains the convex polyhedron $\Omega(b)$.

For a vector c and a matrix A we denote by c' and A' their respective transpose. We also use both notations $c'x$ and $\langle c, x\rangle$ for the usual scalar product of two vector c and x. We assume that both $A \in \mathbb{R}^{m\times n}$ and $b \in \mathbb{R}^m$ have rational entries.

2.1. Duality for $\mathbb{P}$. It is well-known that the standard duality for (1.2) is obtained from the *Legendre-Fenchel* transform $F(.,c) : \mathbb{R}^m \to \mathbb{R}$ of the value function $b \mapsto f(b, c)$ w.r.t. b, i.e., here (as $y \mapsto f(y, c)$ is concave)

$$\lambda \mapsto F(\lambda, c) := \inf_{y\in\mathbb{R}^m} \langle \lambda, y\rangle - f(y, c), \tag{2.3}$$

which yields the usual dual LP problem

$$\mathbb{P}^* \to \quad \inf_{\lambda\in\mathbb{R}^m} \langle \lambda, b\rangle - F(\lambda, c) = \min_{\lambda\in\mathbb{R}^m} \{b'\lambda \,|\, A'\lambda \geq c\} \tag{2.4}$$

already defined in (1.3).

2.2. Duality for integration. Similarly, the analogue for integration of the Fenchel transform is the two-sided *Laplace* transform $\widehat{F}(.,c) : \mathbb{C}^m \to \mathbb{C}$ of the function $b \mapsto \widehat{f}(b, c)$, given by

$$\lambda \mapsto \widehat{F}(\lambda, c) := \int_{\mathbb{R}^m} \mathrm{e}^{-\langle \lambda, y\rangle} \widehat{f}(y, c)\, dy. \tag{2.5}$$

It turns out that developing (2.5) yields

$$\widehat{F}(\lambda, c) = \prod_{k=1}^{n} \frac{1}{(A'\lambda - c)_k} \qquad [\text{whenever } \Re(A'\lambda - c) > 0]. \tag{2.6}$$

(See e.g. [**6**, p. 798] or [**16**]). Thus, $\widehat{F}(\lambda, c)$ is well-defined provided that

$$\Re(A'\lambda - c) > 0, \tag{2.7}$$

and $\widehat{f}(b,c)$ can be computed by solving the *inverse* Laplace transform problem, that we call the (integration) *dual* problem $\mathbb{I}^*$ of (2.8), that is,

$$\begin{aligned}\mathbb{I}^* \to \widehat{f}(b,c) \quad &:= \quad \frac{1}{(2i\pi)^m}\int_{\gamma-i\infty}^{\gamma+i\infty} \mathrm{e}^{\langle b,\lambda\rangle}\,\widehat{F}(\lambda,c)\,d\lambda \\ (2.8) \qquad\qquad &= \quad \frac{1}{(2i\pi)^m}\int_{\gamma-i\infty}^{\gamma+i\infty} \frac{\mathrm{e}^{\langle b,\lambda\rangle}}{\displaystyle\prod_{k=1}^{n}(A'\lambda-c)_k}\,d\lambda\end{aligned}$$

where $\gamma \in \mathbb{R}^m$ is fixed and satisfies $A'\gamma - c > 0$. Incidently, observe that the domain of definition (2.7) of $\widehat{F}(.,c)$ (projected on $\mathbb{R}^m$) is precisely the interior of the feasible set of the dual problem $\mathbb{P}^*$ in (2.4). We will comment more on this and the link with the logarithmic barrier function for linear programming (see §2.5 below).

We may indeed call $\mathbb{I}^*$ a *dual* problem of $\mathbb{I}$ as it is defined on the space $\mathbb{C}^m$ of variables $\{\lambda_k\}$ associated with the nontrivial constraints $Ax = b$; notice that we also retrieve the standard "ingredients" of the dual optimization problem $\mathbb{P}^*$, namely, $b'\lambda$ and $A'\lambda - c$.

2.3. Comparing $\mathbb{P}, \mathbb{P}^*$ and $\mathbb{I}, \mathbb{I}^*$. One may compute directly $\widehat{f}(b,c)$ by using Cauchy residue techniques. That is, one computes the integral (2.8) by successive one-dimensional complex integrals w.r.t. one variable λ_k at a time (e.g. starting with $\lambda_1, \lambda_2, \ldots$) and by repeated application of Cauchy 's Residue Theorem [**7**]. This is possible because the integrand is a rational fraction, and after application of Cauchy 's Residue Theorem at step k w.r.t. λ_k, the output is still a rational fraction of the remaining variables $\lambda_{k+1}, \ldots, \lambda_m$. For more details the reader is referred to Lasserre and Zeron [**16**]. It is not difficult to see that the whole procedure is a summation of partial results, each of them corresponding to a (multi-pole) vector $\widehat{\lambda} \in \mathbb{R}^m$ that annihilates m terms of n products in the denominator of the integrand.

This is formalized in the nice formula of Brion and Vergne [**6**, Proposition 3.3 p. 820] that we describe below. For the interested reader, there are several other nice closed form formula for $\widehat{f}(b,c)$ notably by Barvinok [**3**], Barvinok and Pommersheim [**4**], and Pukhlikov and Khovanskii [**13**].

2.4. The continuous Brion and Vergne' s formula. The material in this section is taken from [**6**]. Write the matrix $A \in \mathbb{R}^{m\times n}$ as $A = [A_1|\ldots|A_n]$ where $A_j \in \mathbb{R}^m$ denotes the j-th column of A for all $j = 1,\ldots,n$. A subset $\sigma = \{j_1,\ldots,j_m\}$ of $\{1,\ldots,n\}$ is called a *basis* if the sequence $\{A_j\}_{j\in\sigma}$ is a basis of $\mathbb{R}^m$. Let $A_\sigma := [A_{j_1}|\ldots|A_{j_m}]$. If $A_\sigma^{-1}b \geq 0$ let $x(\sigma) \in \mathbb{R}^n$ be the vector such that $x_j = (A_\sigma^{-1}b)_j$ for all $j \in \sigma$ and $x_k = 0$ for all $k \notin \sigma$, i.e., $x(\sigma)$ is a *vertex* of $\Omega(b)$. For each basis σ let $\pi^\sigma \in \mathbb{R}^m$ be the row vector that solves $\pi^\sigma A_j = c_j$ for all $j \in \sigma$. The vector $c \in \mathbb{R}^n$ is said to be *regular* if $c_k - \pi^\sigma A_k \neq 0$ for all $k \notin \sigma$. Denote by V the subspace $\{x \in \mathbb{R}^n \mid Ax = 0\}$. In the Linear Programming terminology, $c_k - \pi^\sigma A_k$ is the so-called *reduced cost* of the variable x_k, with respect to the basis $\{A_j\}_{j\in\sigma}$.

Let $c \in \mathbb{R}^n$ be regular with $-c$ in the interior of the dual cone $(\mathbb{R}^n_+ \cap V)^*$ (which is the case if $A'u > c$ for some $u \in \mathbb{R}^m$). Brion and Vergne' s formula [**6**,

Proposition 3.3, p. 820] states that

$$\widehat{f}(b,c) \;=\; \sum_{x(\sigma):\ \text{vertex of } \Omega(b)} \frac{\mathrm{e}^{\langle c,x(\sigma)\rangle}}{|\det(A_\sigma)|\prod_{k\notin\sigma}(-c_k+\pi^\sigma A_k)}. \tag{2.9}$$

Thus, $\widehat{f}(b,c)$ is a weighted *summation* over the vertices of $\Omega(b)$ whereas $f(b,c)$ is a *maximization* over the vertices (or a summation with $\oplus \equiv \max$).

So, if c is replaced with rc and $x(\sigma^*)$ denotes the vertex of $\Omega(b)$ at which $c'x$ is maximized, we obtain

$$\widehat{f}(b,rc)^{1/r} \;=\; \mathrm{e}^{\langle c,x(\sigma^*)\rangle}\left[\sum_{x(\sigma):\ \text{vertex of } \Omega(b)} \frac{\mathrm{e}^{r\langle c,x(\sigma)-x(\sigma^*)\rangle}}{r^{n-m}|\det(A_\sigma)|\prod_{k\notin\sigma}(-c_k+\pi^\sigma A_k)}\right]^{1/r},$$

from which it easily follows that

$$\lim_{r\to\infty} \ln \widehat{f}(b,rc)^{1/r} \;=\; \langle c,x(\sigma^*)\rangle \;=\; \max_{x\in\Omega(b)} \langle c,x\rangle \;=\; f(b,c),$$

as indicated in (1.6).

2.5. The logarithmic barrier function. It is also worth noticing that

$$\begin{aligned}\widehat{f}(b,rc) \;&=\; \frac{1}{(2i\pi)^m}\int_{\gamma_r-i\infty}^{\gamma_r+i\infty} \frac{\mathrm{e}^{\langle b,\lambda\rangle}}{\prod_{k=1}^{n}(A'\lambda-rc)_k}\,d\lambda\\ &=\; \frac{1}{(2i\pi)^m}\int_{\gamma-i\infty}^{\gamma+i\infty} \frac{r^{m-n}\mathrm{e}^{\langle rb,\lambda\rangle}}{\prod_{k=1}^{n}(A'\lambda-c)_k}\,d\lambda\end{aligned}$$

with $\gamma_r = r\gamma$ and we can see that (up to the constant $(m-n)\ln r$) the logarithm of the integrand is nothing less than the well-known *logarithmic barrier function*

$$\lambda \mapsto \phi_\mu(\lambda,b) \;=\; \mu^{-1}b'\lambda - \sum_{j=1}^{n} \ln\,(A'\lambda-c)_j,$$

with parameter $\mu := 1/r$, of the dual problem $\mathbb{P}^*$ (see e.g. Den Hertog [**9**]). This should not be a surprise as a self-concordant barrier function $x \mapsto \phi_K(x)$ of a cone $K \subset \mathbb{R}^n$ is given by the logarithm of Laplace transform $\int_{K^*} \mathrm{e}^{-\langle x,s\rangle} ds$ of its dual cone K^* (see e.g. Güler [**11**], Truong and Tunçel [**27**]).

Thus, when $r\to\infty$, minimizing the exponential logarithmic barrier function on its domain in $\mathbb{R}^m$ yields the same result as taking its residues.

2.6. Summary. The parallel between $\mathbb{P}, \mathbb{P}^*$ and $\mathbb{I}, \mathbb{I}^*$ is summarized below.

Fenchel-duality:	**Laplace-duality**
$f(b,c) := \max_{Ax=b;\, x\geq 0} c'x$	$\widehat{f}(b,c) := \int_{Ax=b;\, x\geq 0} e^{c'x}\, ds$
$F(\lambda,c) := \inf_{y\in\mathbb{R}^m} \{\lambda' y - f(y,c)\}$	$\widehat{F}(\lambda,c) := \int_{\mathbb{R}^m} e^{-\lambda' y}\, \widehat{f}(y,c)\, dy = \dfrac{1}{\prod_{k=1}^{n}(A'\lambda - c)_k}$
with : $A'\lambda - c \geq 0$	with : $\Re(A'\lambda - c) > 0$
$f(b,c) = \min_{\lambda\in\mathbb{R}^m} \{\lambda' b - F(\lambda,c)\}$ $= \min_{\lambda\in\mathbb{R}^m} \{b'\lambda \mid A'\lambda \geq c\}$	$\widehat{f}(b,c) = \dfrac{1}{(2i\pi)^m} \int_\Gamma e^{\lambda' b} \widehat{F}(\lambda,c)\, d\lambda$ $= \dfrac{1}{(2i\pi)^m} \int_\Gamma \dfrac{e^{\lambda' b}}{\prod_{k=1}^{n}(A'\lambda - c)_k} d\lambda$
Simplex algorithm$\rightarrow$ vertices of $\Omega(b)$. $\rightarrow \max c'x$ over vertices.	Cauchy' s Residue$\rightarrow$ poles of $\widehat{F}(\lambda,c)$. $\rightarrow \sum e^{c'x}$ over vertices.

3. Duality for the discrete problems $\mathbb{I}_d$ and $\mathbb{P}_d$

In the respective *discrete* analogues $\mathbb{P}_d$ and $\mathbb{I}_d$ of (1.2) and (2.2), one replaces the positive cone $\mathbb{R}^n_+$ by $\mathbb{N}^n$, that is, (1.2) becomes the integer program $\mathbb{P}_d$ in (1.1), whereas (2.2) becomes a summation over $\mathbb{N}^n \cap \Omega(b)$, i.e.,

$$\mathbb{I}_d : \quad \widehat{f}_d(b,c) := \sum \{e^{c'x} \mid Ax = b; \quad x \in \mathbb{N}^n\}. \tag{3.1}$$

We here assume that $A \in \mathbb{Z}^{m\times n}$ and $b \in \mathbb{Z}^m$, which implies in particular that the *lattice* $\Lambda := A(\mathbb{Z}^n)$ is a sublattice of $\mathbb{Z}^m$ ($\Lambda \subseteq \mathbb{Z}^m$). Note that b in (1.1) and (3.1) is necessarily in Λ.

In this section we are concerned with what we call the "dual" problem $\mathbb{I}^*_d$ of $\mathbb{I}_d$, the discrete analogue of the dual $\mathbb{I}^*$ of $\mathbb{I}$, and its link with the discrete optimization problem $\mathbb{P}_d$.

3.1. The $\mathbb{Z}$-transform. The natural discrete analogue of the Laplace transform is the so-called $\mathbb{Z}$-transform. Therefore, with $\widehat{f}_d(b,c)$ we associate its (two-sided) $\mathbb{Z}$-transform $\widehat{F}_d(.,c) : \mathbb{C}^m \rightarrow \mathbb{C}$ defined by

$$z \mapsto \widehat{F}_d(z,c) := \sum_{y\in\mathbb{Z}^m} z^{-y} \widehat{f}_d(y,c), \tag{3.2}$$

where the notation z^y with $y \in \mathbb{Z}^m$ stands for $z_1^{y_1} \cdots z_m^{y_m}$. Applying this definition yields

$$\begin{aligned}
\widehat{F}_d(z,c) &= \sum_{y\in\mathbb{Z}^m} z^{-y}\widehat{f}_d(y,c) \\
&= \sum_{y\in\mathbb{Z}^m} z^{-y}\left[\sum_{x\in\mathbb{N}^n;\,Ax=y} e^{c'x}\right] = \sum_{x\in\mathbb{N}^n} e^{c'x}\left[\sum_{y=Ax} z_1^{-y_1}\cdots z_m^{-y_m}\right] \\
&= \sum_{x\in\mathbb{N}^n} e^{c'x} z_1^{-(Ax)_1}\cdots z_m^{-(Ax)_m} \\
&= \prod_{k=1}^n \frac{1}{(1-e^{c_k}z_1^{-A_{1k}}z_2^{-A_{2k}}\cdots z_m^{-A_{mk}})} = \prod_{k=1}^n \frac{1}{(1-e^{c_k}z^{-A_k})}, \qquad (3.3)
\end{aligned}$$

which is well-defined provided that

$$|z_1^{A_{1k}}\cdots z_m^{A_{mk}}|\ (=|z^{A_k}|) > e^{c_k} \quad \forall k=1,\ldots,n. \tag{3.4}$$

Observe that the domain of definition (3.4) of $\widehat{F}_d(.,c)$ is the *exponential* version of (2.7) for $\widehat{F}(.,c)$. Indeed, taking the real part of the logarithm in (3.4) yields (2.7).

3.2. The dual problem $\mathbb{I}_d^*$. Therefore, the value $\widehat{f}_d(b,c)$ is obtained by solving the *inverse* $\mathbb{Z}$-transform problem $\mathbb{I}_d^*$ (that we call the *dual* of $\mathbb{I}_d$)

$$\widehat{f}_d(b,c) = \frac{1}{(2i\pi)^m}\int_{|z_1|=\gamma_1}\cdots\int_{|z_m|=\gamma_m} \widehat{F}_d(z)\, z^{b-e_m}\, dz_m\cdots dz_1, \tag{3.5}$$

where e_m is the unit vector of $\mathbb{R}^m$ and $\gamma \in \mathbb{R}^m$ is a (fixed) vector that satisfies $\gamma_1^{A_{1k}}\gamma_2^{A_{2k}}\cdots\gamma_m^{A_{mk}} > e^{c_k}$ for all $k=1,\ldots,n$. We may indeed call $\mathbb{I}_d^*$ the *dual problem* of $\mathbb{I}_d$ as it is defined on the space $\mathbb{Z}^m$ of dual variables z_k associated with the nontrivial constraints $Ax=b$ of the primal problem $\mathbb{I}_d$. And we have the parallel

Continuous Laplace-duality	**Discrete $\mathbb{Z}$-duality**
$\widehat{f}(b,c) := \int_{Ax=b;\,x\in\mathbb{R}_+^n} e^{c'x}\,ds$	$\widehat{f}_d(b,c) := \sum_{Ax=b;\,x\in\mathbb{N}^n} e^{c'x}$
$\widehat{F}(\lambda,c) := \int_{\mathbb{R}^m} e^{-\lambda'y}\,\widehat{f}(y,c)dy$	$\widehat{F}_d(z,c) := \sum_{y\in\mathbb{Z}^m} z^{-y}\,\widehat{f}_d(y,c)$
$= \prod_{k=1}^n \frac{1}{(A'\lambda-c)_k}$	$= \prod_{k=1}^n \frac{1}{1-e^{c_k}z^{-A_k}}$
with $\Re(A'\lambda-c)>0$.	with $\lvert z^{A_k}\rvert > e^{c_k}, \quad k=1,\ldots,n.$

3.3. Comparing $\mathbb{I}^*$ and $\mathbb{I}_d^*$. Observe that the dual problem $\mathbb{I}_d^*$ in (3.5) is of the same nature as $\mathbb{I}^*$ in (2.8) because both reduce to computing a complex integral whose integrand is a rational function. In particular, as $\mathbb{I}^*$, the problem $\mathbb{I}_d^*$ can be solved by Cauchy residue techniques (see e.g. [**17**]).

However, there is an important difference between $\mathbb{I}^*$ and $\mathbb{I}_d^*$. While the data $\{A_{jk}\}$ appear in $\mathbb{I}^*$ as *coefficients* of the dual variables λ_k in $\widehat{F}(\lambda,c)$, they now appear as *exponents* of the dual variables z_k in $\widehat{F}_d(z,c)$. As an immediate consequence, the rational function $\widehat{F}_d(.,c)$ has many more poles than $\widehat{F}(.,c)$ (by considering one variable at a time), and in particular, many of them are *complex*, whereas $\widehat{F}(.,c)$ has only *real* poles. As a result, the integration of $\widehat{F}_d(z,c)$ is more complicated than that of $\widehat{F}(\lambda,c)$, which is reflected in the *discrete* (or *periodic*) Brion and Vergne' s formula described below. However, we will see that the poles of $\widehat{F}_d(z,c)$ are simply related to those of $\widehat{F}(\lambda,c)$.

3.4. The "discrete" Brion and Vergne' s formula. With same notation used in §2.4, for a basis σ, let $\mu(\sigma)$ be the volume of the convex polytope $\{\sum_{j\in\sigma} t_j A_j \mid 0 \leq t_j \leq 1\}$, normalized so that $\mathrm{vol}\,(\mathbb{R}^m/\Lambda) = 1$ (recall that Λ is the sublattice $A(\mathbb{Z}^n)$ of $\mathbb{Z}^m$). Let $c \in \mathbb{R}^n$ be regular with $-c$ in the interior of $(\mathbb{R}^n_+ \cap V)^*$.

Then if $b \in \Lambda$ is such that $\Omega(b)$ is a simple polyhedron, Brion and Vergne [**6**, p. 821] show that

$$\widehat{f}_d(b,c) = \sum_{x(\sigma):\ \text{vertex of } \Omega(b)} \frac{e^{c'x(\sigma)}}{\mu(\sigma)} U_\sigma(b,c), \tag{3.6}$$

with

$$U_\sigma(b,c) = \sum_{g\in G(\sigma)} \frac{e^{2i\pi b}(g)}{\prod_{k\notin\sigma}(1 - e^{-2i\pi A_k}(g)e^{c_k - \pi^\sigma A_k})}, \tag{3.7}$$

where $G(\sigma) := (\oplus_{j\in\sigma}\mathbb{Z}A_j)^*/\Lambda^*$ (where * denotes the dual lattice); it is a finite abelian group of order $\mu(\sigma)$, and with (finitely many) characters $e^{2i\pi b}$ for all $b \in \Lambda$; in particular, writing $A_k = \sum_{j\in\sigma} u_{jk}A_j$ for all $k \notin \sigma$,

$$e^{2i\pi A_k}(g) = e^{2i\pi \sum_{j\in\sigma} u_{jk} g_j} \quad k \notin \sigma.$$

Formula (3.6) is the *discrete* (or, *periodic*) analogue of (2.9). Again and as in (2.9), notice the importance of the *reduced-cost* $c_k - \pi^\sigma A_k$ in the expression of $\widehat{f}_d(b,c)$.

3.5. The discrete optimization problem $\mathbb{P}_d$. We now are in position to see how $\mathbb{I}_d^*$ provides some nice information about the optimal value $f_d(b,c)$ of the discrete optimization problem $\mathbb{P}_d$.

THEOREM 3.1 (Lasserre [**20**]). *Let $A \in \mathbb{Z}^{m\times n}, b \in \mathbb{Z}^m$ and let $c \in \mathbb{Z}^n$ be regular with $-c$ in the interior of $(\mathbb{R}^n_+ \cap V)^*$. With $b \in A(\mathbb{Z}^n)$ let $\Omega(b)$ be a simple polyhedron, and let $q \in \mathbb{N}$ be the least common multiple of $\{\mu(\sigma)\}_{\text{bases}\,\sigma}$*

If $Ax = b$ has no solution $x \in \mathbb{N}^n$ then $f_d(b,c) = -\infty$, else assume that

$$\max_{x(\sigma):\ \text{vertex of } \Omega(b)} \left[c'x(\sigma) + \lim_{r\to\infty} \frac{1}{r} \ln U_\sigma(b, rc)\right], \tag{3.8}$$

is attained at a unique vertex $x(\sigma^*)$ of $\Omega(b)$. Then $x(\sigma^*)$ is the optimal vertex of the linear program $\mathbb{P}$, and

$$\begin{aligned} f_d(b,c) &= c'x(\sigma^*) + \lim_{r\to\infty} \frac{1}{r} \ln U_{\sigma^*}(b, rc) \\ &= c'x(\sigma^*) + \frac{1}{q}(\deg(P_{\sigma^* b}) - \deg(Q_{\sigma^* b})) \end{aligned} \tag{3.9}$$

for some real-valued univariate polynomials $P_{\sigma^* b}, Q_{\sigma^* b}$.

Moreover, the term $\lim_{r\to\infty} \ln U_{\sigma^*}(b, rc)/r$ or $(\deg(P_{\sigma^* b}) - \deg(Q_{\sigma^* b}))/q$ in (3.8), is a sum of some reduced costs $c_k - \pi^{\sigma^*} A_k$ (with $k \notin \sigma^*$).

REMARK 3.2. Of course, (3.9) is not easy to obtain but it shows that the optimal value $f_d(b,c)$ of $\mathbb{P}_d$ is strongly related to the various complex poles of $\widehat{F}_d(z,c)$. It is also interesting to note the crucial role played by the reduced costs $(c_k - \pi^\sigma A_k)$ in linear programming. Indeed, under the condition of uniqueness of the "max" in (3.8), the optimal value $f_d(b,c)$ is the value of $c'x$ at the optimal vertex $x(\sigma^*)$ of the linear program $\mathbb{P}$, plus a sum of some reduced costs. Thus, as for the LP problem $\mathbb{P}$, the optimal value $f_d(b,c)$ of $\mathbb{P}_d$ can be found by inspection of (some sums of) reduced costs associated with each vertex of $\Omega(b)$. In fact, (3.9) also means that the *Gomory relaxation* introduced in Gomory [**10**] is *exact*. For more details on the link between Brion and Vergne' s formula and Gomory relaxations, the interested reader is referred to Lasserre [**20**].

We next derive an asymptotic result that relates the respective optimal values $f_d(b,c)$ and $f(b,c)$ of $\mathbb{P}_d$ and $\mathbb{P}$.

COROLLARY 3.3 (Lasserre [**18**]). *Let $A \in \mathbb{Z}^{m\times n}, b \in \mathbb{Z}^m$ and let $c \in \mathbb{R}^n$ be regular with $-c$ in the interior of $(\mathbb{R}^n_+ \cap V)^*$. With $b \in A(\mathbb{Z}^n)$ let $\Omega(b)$ be a simple polyhedron and let $x^* \in \Omega(b)$ be an optimal vertex of $\mathbb{P}$, that is, $f(b,c) = c'x^* = c'x(\sigma^*)$ with σ^* the unique optimal basis of $\mathbb{P}$. Then for $t \in \mathbb{N}$ sufficiently large,*

$$f_d(tb,c) - f(tb,c) \;=\; \lim_{r\to\infty}\left[\frac{1}{r}\ln U_{\sigma^*}(tb,rc)\right]. \tag{3.10}$$

In particular, for $t \in \mathbb{N}$ sufficently large, the function $t \mapsto f(tb,c) - f_d(tb,c)$ is periodic (constant) with period $\mu(\sigma^*)$.

Thus, when b in Corollary 3.3 is sufficiently large, say $b = tb_0$ with $b_0 \in \Lambda$ and $t \in \mathbb{N}$, the "max" in (3.8) is attained at the unique optimal basis σ^* of the LP (1.2). Equivalently, the Gomory relaxation is exact for large values of b. This result and the periodicity property was already proved in Gomory [**10**]. ¿From Remark 3.2 it also follows that for sufficiently large $t \in \mathbb{N}$, the optimal value $f_d(tb,c)$ is equal to $f(tb,c)$ plus a sum of some *reduced costs* $c_k - \pi^{\sigma^*}A_k$ (with $k \notin \sigma^*$) with respect to the optimal basis σ^*.

3.6. A dual comparison of $\mathbb{P}$ and $\mathbb{P}_d$. We now provide an alternative formulation of Brion and Vergne 's discrete formula (3.6), which explicitly relates dual variables of $\mathbb{P}$ and $\mathbb{P}_d$. Recall that a *feasible basis* of the linear program $\mathbb{P}$ is a basis σ such that $A_\sigma^{-1}b \geq 0$. Thus, let σ be a feasible basis of the linear program $\mathbb{P}$, and consider the system of m equations in $\mathbb{C}^m$:

$$z_1^{A_{1j}}\cdots z_m^{A_{mj}} \;=\; e^{c_j} \qquad j \in \sigma. \tag{3.11}$$

Recall that A_σ is the nonsingular matrix $[A_{j_1}|\cdots|A_{j_m}]$, with $j_k \in \sigma$ for all $k = 1,\ldots,m$. The above system (3.11) has $\rho(\sigma)$ $(= \det(A_\sigma))$ distinct solutions $\{z(k)\}_{k=1}^{\rho(\sigma)}$ in $\mathbb{C}^m$, written

$$z(k) \;=\; e^{\lambda}e^{2i\pi\theta(k)} \qquad k = 1,\ldots,\rho(\sigma), \tag{3.12}$$

for some $\lambda \in \mathbb{R}^m$, and $\rho(\sigma)$ vectors $\{\theta(k)\}$ in $\mathbb{R}^m$.

Indeed, writing $z = e^{\lambda}e^{2i\pi\theta}$ (i.e., the vector $\{e^{\lambda_j}e^{2i\pi\theta_j}\}_{j=1}^m$ in $\mathbb{C}^m$), and passing to the logarithm in (3.11), yields

$$A'_\sigma\lambda + 2i\pi\, A'_\sigma\theta \;=\; c_\sigma \tag{3.13}$$

where $c_\sigma \in \mathbb{R}^m$ is the vector $\{c_j\}_{j\in\sigma}$. Thus, $\lambda \in \mathbb{R}^m$ is the unique solution of $A'_\sigma\lambda = c_\sigma$ and θ satisfies

$$A'_\sigma\theta \in \mathbb{Z}^m. \tag{3.14}$$

Equivalently, θ belongs to $(\oplus_{j\in\sigma}A_j\mathbb{Z})^*$, the dual lattice of $\oplus_{j\in\sigma}A_j\mathbb{Z}$.

Thus, there is a one-to-one correspondence between the $\rho(\sigma)$ solutions $\{\theta(k)\}$ and the finite group $G'(\sigma) = (\oplus_{j\in\sigma}A_j\mathbb{Z})^*/\mathbb{Z}^m$, and $G(\sigma)$ is a subgroup of $G'(\sigma)$. Thus, with $G(\sigma) = \{g_1,\ldots,g_s\}$ and $s := \mu(\sigma)$, we can write $(A'_\sigma)^{-1}g_k = \theta_{g_k} = \theta(k)$, so that for every character $e^{2i\pi y}$ of $G(\sigma)$, $y \in \Lambda$, we have

$$e^{2i\pi y}(g) = e^{2i\pi y'\theta_g} \qquad y \in \Lambda,\ g \in G(\sigma). \tag{3.15}$$

and

$$e^{2i\pi A_j}(g) = e^{2i\pi A'_j\theta_g} = 1 \qquad j \in \sigma. \tag{3.16}$$

For every feasible basis σ, denote by $\{z_g\}_{g\in G(\sigma)}$ these $\mu(\sigma)$ solutions of (3.12), that is,

$$z_g = e^\lambda e^{2i\pi\theta_g} \in \mathbb{C}^m, \qquad g \in G(\sigma), \tag{3.17}$$

with $\lambda = (A'_\sigma)^{-1}c_\sigma$, and where $e^\lambda \in \mathbb{R}^m$ is the vector $\{e^{\lambda_i}\}_{i=1}^m$.

So, in the linear program $\mathbb{P}$ we have a dual vector $\lambda \in \mathbb{R}^m$ associated with each basis σ. In the integer program $\mathbb{P}_d$, with each (same) basis σ are now associated $\mu(\sigma)$ "dual" (complex) vectors $\lambda + 2i\pi\theta_g$, $g \in G(\sigma)$. Hence, with a basis σ in linear programming, the "dual variables" in integer programming are obtained from (a), the corrresponding dual variables $\lambda \in \mathbb{R}^m$ in linear programming, and (b), a periodic correction term $2i\pi\theta_g \in \mathbb{C}^m$, $g \in G(\sigma)$.

We next introduce what we call the *vertex residue function.*

DEFINITION 3.4. Let $b \in \Lambda$ and let $c \in \mathbb{R}^n$ be regular. Let σ be a feasible basis of the linear program $\mathbb{P}$ and for every $r \in \mathbb{N}$, let $\{z_{gr}\}_{g\in G(\sigma)}$ be as in (3.17), with rc instead of c, that is,

$$z_{gr} = e^{r\lambda}e^{2i\pi\theta_g} \in \mathbb{C}^m; \qquad g \in G(\sigma), \quad (\text{with } \lambda = (A'_\sigma)^{-1}c_\sigma).$$

The vertex residue function associated with a feasible basis σ of the linear program $\mathbb{P}$, is the function $R_\sigma(z_g,.) : \mathbb{N}\rightarrow\mathbb{R}$ defined by :

$$r \mapsto R_\sigma(z_g,r) := \frac{1}{\mu(\sigma)}\sum_{g\in G(\sigma)}\frac{z_{gr}^b}{\prod_{k\notin\sigma}(1 - z_{gr}^{-A_k}e^{rc_k})}, \tag{3.18}$$

which is well-defined because when c is regular, $|z_{gr}|^{A_k} \neq e^{rc_k}$ for all $k \notin \sigma$.

The name *vertex residue* is now clear because in the integration (3.5), $R_\sigma(z_g,r)$ is to be interpreted as a generalized Cauchy residue, with respect to the $\mu(\sigma)$ "poles" $\{z_{gr}\}$ of the generating function $z \mapsto \widehat{F}_d(z,rc)$.

Recall that from Corollary 3.3, when $b \in \Lambda$ is sufficiently large, say $b = tb_0$ with $b_0 \in \Lambda$ and some large $t \in \mathbb{N}$, the "max" in (3.8) is attained at the unique optimal basis σ^* of the linear program $\mathbb{P}$.

PROPOSITION 3.5 (Lasserre [**20**]). *Let c be regular with $-c \in (\mathbb{R}^n_+ \cap V)^*$, and let $b \in \Lambda$ with $\Omega(b)$ being a simple polyhedron. Assume that b is sufficiently large* *so that the max in (3.8) is attained at the unique optimal basis* σ^* *of the linear program* $\mathbb{P}$. *Let* $\{z_g\}_{g\in G(\sigma^*)}$ *be as in (3.17) with* $\sigma = \sigma^*$.

Then the optimal value of $\mathbb{P}_d$ *satisfies*

$$f_d(b,c) = \lim_{r\to\infty} \frac{1}{r} \ln \left[\frac{1}{\mu(\sigma^*)} \sum_{g\in G(\sigma^*)} \frac{z_{gr}^b}{\prod_{k\notin\sigma^*}(1 - z_{gr}^{-A_k} e^{rc_k})} \right]$$

$$(3.19) \qquad = \lim_{r\to\infty} \frac{1}{r} \ln R_{\sigma^*}(z_g, r),$$

and the optimal value of $\mathbb{P}$ *satisfies*

$$f(b,c) = \lim_{r\to\infty} \frac{1}{r} \ln \left[\frac{1}{\mu(\sigma^*)} \sum_{g\in G(\sigma^*)} \frac{|z_{gr}|^b}{\prod_{k\notin\sigma^*}(1 - |z_{gr}|^{-A_k} e^{rc_k})} \right]$$

$$(3.20) \qquad = \lim_{r\to\infty} \frac{1}{r} \ln R_{\sigma^*}(|z_g|, r).$$

Proposition 3.5 shows that there is indeed a strong relationship between the integer program $\mathbb{P}_d$ and its continuous analogue, the linear program $\mathbb{P}$. Both optimal values obey exactly the same formula (3.19), but for the continuous version, the complex vector $z_g \in \mathbb{C}^m$ is replaced with the vector $|z_g| = e^{\lambda^*} \in \mathbb{R}^m$ of its component moduli, where $\lambda^* \in \mathbb{R}^m$ is the optimal solution of the LP dual of $\mathbb{P}$. In summary, when $c \in \mathbb{R}^n$ is regular and $b \in \gamma \cap \Lambda$ is sufficiently large, we have the correspondence

Linear program $\mathbb{P}$	**Integer program** $\mathbb{P}_d$
unique optimal basis σ^*	unique optimal basis σ^*
1 optimal dual vector $\lambda^* \in \mathbb{R}^m$	$\mu(\sigma^*)$ dual vectors $z_g \in \mathbb{C}^m,\ g \in G(\sigma^*)$
	$\ln z_g = \lambda^* + 2i\pi\theta_g$
$f(b,c) = \lim_{r\to\infty} \frac{1}{r} \ln R_{\sigma^*}(\lvert z_g\rvert, r)$	$f_d(b,c) = \lim_{r\to\infty} \frac{1}{r} \ln R_{\sigma^*}(z_g, r)$

4. A discrete Farkas lemma

In this section we are interested in a discrete analogue of the continuous Farkas lemma. That is, with $A \in \mathbb{Z}^{m\times n}, b \in \mathbb{Z}^m$, consider the issue of existence of a *nonnegative integral* solution $x \in \mathbb{N}^n$ to the system of linear equations $Ax = b$.

The (continuous) Farkas Lemma, which states that given $A \in \mathbb{R}^{m\times n}, b \in \mathbb{R}^m$,

$$(4.1) \qquad \{x \in \mathbb{R}^n \mid Ax = b,\ x \geq 0\} \neq \emptyset \Leftrightarrow [A'\lambda \geq 0] \Rightarrow b'\lambda \geq 0,$$

has no discrete analogue in an *explicit* form. For instance, the *Gomory functions* used in Blair and Jeroslow [**5**] (see also Schrijver [**26**, Corollary 23.4b]) are implicitly and iteratively defined, and not directly in terms of the data A, b. On the other hand, for various characterizations of feasibility of linear diophantine equations $Ax = b,\ x \in \mathbb{Z}^n$, the interested reader is referred to Schrijver [**26**, §4].

Before proceeding to the general case $A \in \mathbb{Z}^{m\times n}$, we first consider the case $A \in \mathbb{N}^{m\times n}$ where A (and thus b) has only nonnegative entries.

THEOREM 4.1 (Lasserre [**19**]). *Let $A \in \mathbb{N}^{m\times n}, b \in \mathbb{N}^m$. Then the following two propositions (i) and (ii) are equivalent :*

(i) *The linear system $Ax = b$ has a solution $x \in \mathbb{N}^n$.*

(ii) *The real-valued polynomial $z \mapsto z^b - 1 := z_1^{b_1} \cdots z_m^{b_m} - 1$ can be written*

$$z^b - 1 = \sum_{j=1}^{n} Q_j(z)(z^{A_j} - 1), \tag{4.2}$$

for some real-valued polynomials $Q_j \in \mathbb{R}[z_1, \ldots, z_m]$, $j = 1, \ldots, n$, all of them with nonnegative coefficients.

In addition, the degree of the Q_j's in (4.2) is bounded by

$$b^* := \sum_{j=1}^{m} b_j - \min_k \sum_{j=1}^{m} A_{jk}. \tag{4.3}$$

Hence Theorem 4.1 reduces the issue of existence of a solution $x \in \mathbb{N}^n$ to a particular *ideal membership problem*, that is, $Ax = b$ has a solution $x \in \mathbb{N}^n$ if and only if the polynomial $z^b - 1$ belongs to the *binomial ideal* $I = \langle z^{A_j} - 1\rangle_{j=1,\ldots,n} \subset \mathbb{R}[z_1, \ldots, z_m]$ for some weights Q_j with *nonnegative* coefficients.

REMARK 4.2. (a) From the proof of Theorem 4.1 in Lasserre [**19**], whenever they exist, the polynomials $Q_j = \sum_{\alpha\in\mathbb{N}^m} Q_{j\alpha} z^\alpha$ may be restricted to have only monomials z^α with $\alpha \leq b - A_j$ for all $j = 1, \ldots, n$. Therefore, checking the existence of such polynomials $\{Q_j\}$ in (4.2), can be done by solving a linear system

$$\mathrm{M}\, q = r, \quad q \geq 0, \tag{4.4}$$

where q is the vector of (nonnegative) coefficients of the polynomials$\{Q_j\}$, and the constraints state that the polynomial $z \mapsto z^b - 1 - \sum_j Q_j(z)(z^{A_j} - 1)$ is identically null. This in turn reduces to solving a LP problem with $s := \sum_{j=1}^n s_j$ variables (with $s_j := \prod_{k=1}^m (b_k - A_{kj} + 1)$, for all $j = 1, \ldots, n$), and $p := \prod_{k=1}^m (b_k + 1)$ equality constraints. In addition, in view of (4.2), this LP has a matrix of constraints $\mathrm{M} \in \mathbb{R}^{p\times s}$, and a right-hand-side $r \in \mathbb{R}^p$, with only 0 and ±1 coefficients.

(b) From the proof of Theorem 4.1, it easily follows that one may even enforce the weights Q_j in (4.2) to be polynomials in $\mathbb{Z}[z_1, \ldots, z_m]$ (instead of $\mathbb{R}[z_1, \ldots, z_m]$) with nonnegative coefficients. However, (a) shows that the strength of Theorem 4.1 is precisely to allow $Q_j \in \mathbb{R}[z_1, \ldots, z_m]$ as it permits to check feasibility by solving a (continuous) linear program. By enforcing $Q_j \in \mathbb{Z}[z_1, \ldots, z_m]$ one would end up with an integer linear system of size larger than that of the original problem.

For the general case $A \in \mathbb{Z}^{m\times n}$, i.e., when A may have negative entries, we assume that the convex polyhedron $\Omega := \{x \in \mathbb{R}^n \mid Ax = b;\ x \geq 0\}$ is *compact*, so that this polyhedron is actually a polytope.

The above arguments cannot be repeated because of the occurence of negative powers. However, let $\alpha \in \mathbb{N}^n, \beta \in \mathbb{N}$ be such that

$$b_j + \beta \geq 0, \quad \text{and} \quad A_{jk} + \alpha_k \geq 0, \quad k = 1, \ldots, n; \quad j = 1, \ldots, m. \tag{4.5}$$

As Ω is compact,

$$\max_{x\in\mathbb{N}^n} \{\sum_{j=1}^n \alpha_j x_j \,|\, Ax = b\} \leq \max_{x\in\mathbb{R}^n_+} \{\sum_{j=1}^n \alpha_j x_j \,|\, Ax = b\} =: \rho^*(\alpha) < \infty. \tag{4.6}$$

Observe that given $\alpha \in \mathbb{N}^n$, the scalar $\rho^*(\alpha)$ is easily calculated by solving a LP problem.

COROLLARY 4.3 (Lasserre [**19**]). *Let $A \in \mathbb{Z}^{m\times n}, b \in \mathbb{Z}^m$ and assume that $\Omega := \{x \in \mathbb{R}^n_+ \,|\, Ax = b\}$ is compact. Let $\alpha \in \mathbb{N}^n$ and $\beta \in \mathbb{N}$ be as in (4.5) with $\beta \geq \rho^*(\alpha)$ (cf. (4.6)). Then the following two propositions (i) and (ii) are equivalent :*

(i) *The system of linear equations $Ax = b$ has a solution $x \in \mathbb{N}^n$.*

(ii) *The real-valued polynomial $z \mapsto z^b(zy)^\beta - 1 \in \mathbb{R}[z_1, \ldots, z_m, y]$ can be written*

$$z^b(zy)^\beta - 1 = Q_0(z,y)(zy-1) + \sum_{j=1}^n Q_j(z,y)(z^{A_j}(zy)^{\alpha_j} - 1) \tag{4.7}$$

for some real-valued polynomials $\{Q_j\}_{j=0}^n$ in $\mathbb{R}[z_1, \ldots, z_m, y]$, all with nonnegative coefficients.

The degree of the Q_j's in (4.7) is bounded by

$$(m+1)\beta + \sum_{j=1}^m b_j - \min\left[m+1, \min_{k=1,\ldots,n}\left[(m+1)\alpha_k + \sum_{j=1}^m A_{jk}\right]\right].$$

Indeed Theorems 4.1, and Corollary 4.3 have the flavor of a Farkas Lemma as it is stated with the transpose A' of A and involves the dual variables z_k associated with the constraints $Ax = b$. In addition, and as expected, it implies the continuous Farkas lemma because if $\{x \in \mathbb{N}^n \,|\, Ax = b\} \neq \emptyset$, then from (4.7), and with $z := \mathrm{e}^\lambda$, and $y := (z_1 \cdots z_m)^{-1}$,

$$\mathrm{e}^{b'\lambda} - 1 = \sum_{j=1}^m Q_j(e^{\lambda_1}, \ldots e^{\lambda_m}, e^{-\sum_i \lambda_i})(\mathrm{e}^{(A'\lambda)_j} - 1). \tag{4.8}$$

Therefore, $A'\lambda \geq 0 \Rightarrow \mathrm{e}^{(A'\lambda)_j} - 1 \geq 0$ for all $j = 1, \ldots, n$, and as the Q_j have nonnegative coefficients, we have $e^{b'\lambda} - 1 \geq 0$ which in turn implies $b'\lambda \geq 0$.

Equivalently, evaluating the partial derivatives of both sides of (4.8) with respect to λ_j, at the point $\lambda = 0$, yields $b_j = \sum_{k=1}^n A_{jk}x_k$ for all $j = 1, \ldots, n$, with $x_k := Q_k(1, \ldots, 1) \geq 0$. Thus, $Ax = b$ for some $x \in \mathbb{R}^n_+$.

5. An equivalent LP

In this section we provide a linear program equivalent to the integer program $\mathbb{P}_d$ in (1.1). For ease of exposition we restrict to the case where $A \in \mathbb{N}^{m\times n}, b \in \mathbb{N}^m$.

From Remark 4.2(a), checking the existence of polynomials $\{Q_j\}_{j=1}^n$ in (4.2) reduces to checking whether the linear system $\mathrm{M}\, q = r$ in (4.4) has a nonnegative solution $q \in \mathbb{R}^s$, where the matrix $\mathrm{M} \in \mathbb{R}^{p\times s}$ and the vector $r \in \mathbb{R}^p$ have only 0 and ± 1 coefficients. The vector $q \in \mathbb{R}^s$ can be written $q' = [q'_1, \ldots, q'_n]$ with $q_j \in \mathbb{R}^{s_j}$ and $s_j = \prod_{k=1}^m (b_k - A_{kj} + 1)$, for all $j = 1, \ldots, n$.

With $\mathrm{e}_{s_j} \in \mathbb{R}^{s_j}$ being a row vector of ones, let $\mathrm{E} \in \mathbb{R}^{p\times s}$ be the block diagonal matrix

$$\mathrm{E} := \begin{bmatrix} \mathrm{e}_{s_1} & 0 & \dots & 0 \\ 0 & \mathrm{e}_{s_2} & 0 & \dots \\ \dots & \dots & \dots & \dots \\ 0 & \dots & 0 & \mathrm{e}_{s_n} \end{bmatrix}. \tag{5.1}$$

Similarly, given $c \in \mathbb{R}^n$, let $\widehat{c} := \mathrm{E}'c$, and consider the LP

$$\mathbb{Q} \quad \rightarrow \quad \max_{q\in\mathbb{R}^s} \{\, \widehat{c}'q \mid \quad \mathrm{M}\,q \;=\; r, \quad q \geq 0\}. \tag{5.2}$$

with optimal value denoted by $\max \mathbb{Q}$.

THEOREM 5.1 (Lasserre [**21**]). *Let $A \in \mathbb{N}^{m\times n}, b \in \mathbb{N}^m$. Then :*
(a) $\max \mathbb{Q} \;=\; f_d(b,c) \;=\; \max\{\, c'x \mid \quad Ax = b; \quad x \in \mathbb{N}^n\}$.
(b) *For every optimal vertex $q^* \in \mathbb{R}^s$ of the linear program $\mathbb{Q}$, the vector $x^* := \mathrm{E}q^*$ is in $\mathbb{N}^n$ and is an optimal solution of $\mathbb{P}_d$.*

Of course, the linear program $\mathbb{Q}$ has a size which is *not* polynomial in the input size of $\mathbb{P}_d$. Theorem 5.1 states that the linear program $\mathbb{Q}$ is *equivalent* to the integer program $\mathbb{P}_d$. With each vertex of the polyhedron $\{q \in \mathbb{R}^s \mid \mathrm{M}q = r,\, q \geq 0\}$ is associated a vector $x \in \mathbb{N}^n$ that satisfies $Ax = b$. For more details the interested reader is referred to Lasserre [**21**].

6. Duality and superadditivity

We now relate the linear program $\mathbb{Q}$ in (5.2) to the abstract dual $\mathbb{P}_d^*$ of $\mathbb{P}_d$, defined in (1.4). For ease of exposition, and as in §5, again we assume that A has nonnegative entries, i.e., $A \in \mathbb{N}^{m\times n}$ (hence $b \in \mathbb{N}^m$).

With $b \in \mathbb{N}^m$, let $\mathcal{D} \subset \mathbb{N}^m$ be the set

$$\mathcal{D} \;:=\; \prod_{j=1}^{m} \{0, 1, \dots, b_j\}. \tag{6.1}$$

In view of the simple form of the matrix M of the linear program $\mathbb{Q}$ in (5.2), its LP dual $\mathbb{Q}^*$ is easy to state. Namely, $\mathbb{Q}^*$ is the LP

$$\mathbb{Q}^* \quad \rightarrow \left\{ \begin{array}{ll} & \min_{\gamma} \gamma(b) - \gamma(0) \\ \text{s.t.} & \gamma(\alpha + A_j) - \gamma(\alpha) \quad \geq c_j, \quad \alpha + A_j \in \mathcal{D};\; j = 1, \dots, n, \end{array} \right. \tag{6.2}$$

with optimal value denoted by $\min \mathbb{Q}^*$.

From Theorem 5.1 and standard LP duality we have

$$f_d(b,c) \;=\; \max \mathbb{Q} \;=\; \min \mathbb{Q}^*. \tag{6.3}$$

Now, let Γ be the set of functions $\pi : \mathbb{N}^m \rightarrow \mathbb{R}$ such that

$$\pi(0) \;=\; 0 \quad \text{and} \quad \left\{ \begin{array}{lll} \pi(\alpha) \text{ is finite} & \text{if} & \alpha \in \mathcal{D} \\ \pi(\alpha) = +\infty & \text{if} & \alpha \notin \mathcal{D} \end{array} \right. \tag{6.4}$$

so that $\mathbb{Q}^*$ has the following equivalent formulation

$$\mathbb{Q}^* \quad \rightarrow \left\{ \begin{array}{ll} \min_{\pi\in\Gamma} & \pi(b) \\ \text{s.t.} & \inf_{\alpha\in\mathcal{D}} \{\pi(\alpha + A_j) - \pi(\alpha)\} \quad \geq c_j, \quad j = 1, \dots, n. \end{array} \right. \tag{6.5}$$

Next, with each $\pi \in \Gamma$ define $f_\pi : \mathbb{N}^m \to \mathbb{R}$ by

$$\beta \mapsto f_\pi(\beta) := \inf_{\alpha \in \mathcal{D}} \pi(\alpha+\beta) - \pi(\alpha).$$

Observe that

$$f_\pi(0) = 0 \quad \text{and} \quad f_\pi(x) \quad \begin{cases} \text{is finite} & \text{if } x \in \mathcal{D} \\ = +\infty & \text{if } x \notin \mathcal{D}, \end{cases}$$

that is $f_\pi \in \Gamma$. In addition, f_π is *superadditive*. Consider the optimization problem

$$(6.6) \qquad \widehat{\mathbb{Q}}^* \quad \to \begin{cases} \min\limits_{\pi \in \Gamma} & f_\pi(b) \\ \text{s.t.} & f_\pi(A_j) \quad \geq c_j, \quad j = 1, \ldots, n, \end{cases}.$$

with associated optimal value denoted by $\min \widehat{\mathbb{Q}}^*$. As $\pi(b) \geq f_\pi(b)$ it follows that $\min \widehat{\mathbb{Q}}^* \leq \min \mathbb{Q}^*$. Moreover, let $x \in \mathbb{N}^n$ satisfy $Ax = b$. Using the superadditivity of f_π and proceeding as in (1.5) (with f_π in lieu of f), we obtain $c'x \leq f_\pi(b)$. Thus, as $x \in \mathbb{N}^n$ was an arbitrary solution of $Ax = b$, it follows that $f_\pi(b) \geq \max \mathbb{Q}$ for every $\pi \in \Gamma$. Hence $\max \mathbb{Q} \leq \min \widehat{\mathbb{Q}}^* \leq \min \mathbb{Q}^*$ and from (6.3) it follows that $\max \mathbb{Q} = \min \widehat{\mathbb{Q}}^*$.

Thus, via its equivalent formulation $\widehat{\mathbb{Q}}^*$, we have related the explicit dual $\mathbb{Q}^*$ of $\mathbb{Q}$, to the abstract dual $\mathbb{P}_d^*$ of the integer program $\mathbb{P}_d$, defined in (1.4). In particular, instead of considering the set Δ of superadditive functions $f : \mathbb{R}^m \to \mathbb{R}$ in (1.4), it is enough to consider the set of superadditive functions $f_\pi : \mathbb{N}^m \to \mathbb{R}$ obtained from $\pi \in \Gamma$. It thus permits to replace the abstract dual $\mathbb{P}_d^*$ with the equivalent linear program (6.2).

7. Conclusion

We have shown that a natural duality for the integer program $\mathbb{P}_d$ can be derived from the use of generating functions associated with the counting version I_d of $\mathbb{P}_d$. On then retrieves analogues of some well-known dual concepts of linear programming duality. We hope that some of the new insights on $\mathbb{P}_d$ provided by this approach will be further exploited to yield more efficient solving procedures.

References

[1] W.W. Adams, P. Loustaunau, *An Introduction to Gröbner Bases*, American Mathematical Society, Providence, RI, 1994.

[2] F. Bacelli, G. Cohen, G.J. Olsder, J-P. Quadrat. *Synchronization and Linearity*, John Wiley & Sons, Chichester, 1992.

[3] A.I. Barvinok, *Computing the volume, couting integral points and exponentials sums*, Discr. Comp. Geom. **10** (1993), 123–141.

[4] A.I. Barvinok, J.E. Pommersheim, *An algorithmic theory of lattice points in polyhedra*, in *New Perspectives in Algebraic Combinatorics*, MSRI Publications **38** (1999), 91–147.

[5] C.E. Blair, R.G. Jeroslow, *The value function of an integer program*, Math. Prog. **23** (1982), 237–273.

[6] M. Brion, M. Vergne, *Residue formulae, vector partition functions and lattice points in rational polytopes*, J. Amer. Math. Soc. **10** (1997), 797–833.

[7] J.B. Conway, *Functions of a complex variable I*, 2nd ed., Springer, New York, 1978.

[8] G. Cornuejols, Y. Li, *Elementary closures for integer programs*, Oper. Res. Letters **28** (2001), 1–8.

[9] D. den Hertog, *Interior Point Approach to Linear, Quadratic and Convex Programming*, Kluwer Academic Publishers, Dordrecht, 1994.

[10] R.E. Gomory, *Some Polyhedra Related to Combinatorial Problems*, Lin. Alg. Appl. **2** (1969), 451–558.

[11] O. Güler, *Barrier functions in interior point methods*, Math. Oper. Res. **21** (1996), 860–885.

[12] A. Iosevich, *Curvature, combinatorics, and the Fourier transform*, Notices of the AMS **48** (2001), 577–583.

[13] A. Pukhlikov, A. Khovanskii. *A Riemann-Roch theorem for integrals and sums of quasipolynomials over virtual polytopes*, St. Petersburg Math. J. **4** (1993), 789–812.

[14] R. Jeroslow, *An introduction to the theory of cutting-planes*, Ann. Discrete Math. **5** (1979), 71–95.

[15] E.L. Johnson, *Integer Programming. Facets, Subadditivity, and Duality for Group and Semigroup Problems*, SIAM, Philadelphia, Pennsylvania, 1980.

[16] J.B. Lasserre, E.S. Zeron, *A Laplace transform algorithm for the volume of a convex polytope*, J. of the ACM **48** (2001), 1126–1140.

[17] J.B. Lasserre, E.S. Zeron, *On counting integral points in a convex polytope*, Math. Oper. Res. **28** (2003), 853–870.

[18] J.B. Lasserre, *La valeur optimale des programmes entiers*, C. R. Acad. Sci. Paris . Ser. I **335** (2002), 863–866.

[19] J.B. Lasserre, *A discrete Farkas lemma*, In : *Computational Science and its Applications, ICCSA 2003, Montreal*, Kumar V, Gavrilova M.L., Tan C.J.K., l' Ecuyer P. (Eds), Lecture Notes in Computer Sciences 2667, Springer, 2003. Also to appear in Discr. Optim.

[20] J.B. Lasserre, *Generating functions and duality for integer programs*, Discr. Optim., to appear.

[21] J.B. Lasserre, *The integer hull of a convex rational polytope*, Discr. Comput. Geom., to appear.

[22] M. Laurent, *A comparison of the Sherali-Adams Lovász-Schrijver and Lasserre relaxations for 0-1 programming*, Math. Oper. Res. **28** (2003), 470–496.

[23] G.L. Litvinov, V.P. Maslov, G.B. Shpiz. *Linear functionals on idempotent spaces: An algebraic approach*, Doklady Math. **58** (1998), 389–391.

[24] D.S. Mitrinović, J. Sándor, B. Crstici. *Hanbook of Number Theory*, Kluwer Academic Publishers, Dordrecht, 1996.

[25] G.L. Nemhauser, L.A. Wolsey, *Integer and Combinatorial Optimization*, Wiley, John & Sons, 1988.

[26] A. Schrijver, *Theory of Linear and Integer Programming*, John Wiley & Sons, Chichester, 1986.

[27] V.A. Truong, L. Tunçel, *Geometry of homogeneous convex cones, duality mapping, and optimal self-concordant barriers*, Research report COOR #2002-15, University of Waterloo, Waterloo, Canada, 2002.

[28] L.A. Wolsey, *Integer programming duality: Price functions and sensitivity analysis*, Math. Prog. **20** (1981), 173–195.

[29] L.A. Wolsey, *Integer Programming*, John Wiley & Sons, Inc., New York, 1998.

LAAS-CNRS, 7 Avenue du Colonel Roche, 31077 Toulouse Cédex 4, France.
E-mail address: `lasserre@laas.fr`

Proceedings of Symposia in Applied Mathematics
Volume **61**, 2004

The Design and Analysis of Approximation Algorithms: Facility Location as a Case Study

David B. Shmoys

ABSTRACT. Approximation algorithms are polynomial-time algorithms for optimization problems that are guaranteed to find near-optimal solutions. The design and analysis of approximation algorithms for *NP*-hard problems has recently been one of the most active areas in optimization. We will present several different approaches to designing approximation algorithms, by focusing on one problem domain in discrete optimization, that of facility location problems.

1. Introduction to approximation algorithms

Most combinatorial optimization problems are *NP*-hard, and hence are unlikely to have algorithms that are guaranteed to find optimal solutions within polynomial time. One of the primary approaches to cope with this intractability is to settle for less than optimality, and to focus on *approximation algorithms* instead. That is, you design a polynomial-time algorithm for your problem that need not always give the optimal solution, but always produces a solution that is nearly optimal, i.e., with objective function value within some performance guarantee. In recent years, one of the most active areas of research within optimization has been in the design and analysis of approximation algorithms.

In this lecture, we will focus on two closely related computational problems, in order to compare and contrast several different algorithmic approaches that have played leading roles in the recent burst of activity in this field. The central concept for this approach is that of a *ρ-approximation algorithm* for an optimization problem: a polynomial-time algorithm that delivers a feasible solution of objective function value within a factor of ρ of the optimal value. The study of approximation algorithms predates the theory of *NP*-completeness. Some early results, such as the proof due to Vizing that a graph always has an edge coloring with at most one more color than the maximum degree, were not even algorithmically stated, but still contain all of the ideas needed to give an interesting approximation algorithm. Graham, in studying a particular scheduling problem, more explicitly stated the goal of analyzing an efficient heuristic from the perspective of its worst-case error

1991 *Mathematics Subject Classification.* Primary 90B80, 90C27; Secondary 68W40, 68W25.
Key words and phrases. combinatorial optimization, design and analysis of algorithms.
The author was supported in part by NSF Grant CCR-9912422.

bound, or more positively stated, its performance guarantee. The seminal work of Johnson, reacting to the recent discovery of the theory of *NP*-completeness, crystallized the perspective that one approach to coping with the intractability of a problem is to design and analyze approximation algorithms. Much of the history of the work in this area over the past forty years is summarized in the textbook of Vazirani [**6**], the volume of surveys edited by Hochbaum [**3**], and the survey of Shmoys [**4**]. Furthermore, there is also recent work that has provided techniques to prove, for some problems, that no such approximation algorithm can exist. This type of result is also discussed in the references above.

Most combinatorial optimization problems can be formulated as integer programming problems; that is, the set of feasible solutions can be described by decision variables restricted to integer values, and constrained to satisfy a collection of linear inequalities and equations, and the objective function to be optimized can also be expressed as a linear function of the decision variables.

For example, consider the following problem, known as the *uncapacitated facility location problem*, which is to be the main focus of this lecture. In this problem, there is a set of locations $\mathcal{F}$ at which we may build a facility (such as a warehouse), where the cost of building at location i is $f_i > 0$, for each $i \in \mathcal{F}$. There is a set $\mathcal{D}$ of client locations (such as stores) that require to be serviced by a facility, and if a client at location j is assigned to a facility at location i, a cost of c_{ij} is incurred. The objective is to determine a set of locations at which to open facilities so as to minimize the total facility opening and client assignment costs. All of the data are assumed to be non-negative, and furthermore, we will assume that all of the locations are located in a common metric space, so that for any $i, j, k \in \mathcal{N} = F \cup D$, $c_{ik} \leq c_{ij} + c_{jk}$; finally, the costs are also symmetric: for any $i, j \in \mathcal{N}$, $c_{ij} = c_{ji}$. Note that we can think of these costs as representing the distances between locations. We have focused on this "metric" special case for a good reason; without this assumption, one can prove that no good approximation algorithms exist.

There are two types of decision variables x_{ij} and y_i, where each variable y_i, $i \in \mathcal{F}$, indicates whether or not a facility is built at location i, and each variable x_{ij} indicates whether or not the client at location j is assigned to a facility at location i, for each $i \in \mathcal{F}$, $j \in \mathcal{D}$:

$$\text{Minimize} \sum_{i \in \mathcal{F}} f_i y_i + \sum_{i \in \mathcal{F}} \sum_{j \in \mathcal{D}} c_{ij} x_{ij} \tag{1.1}$$

subject to

$$\sum_{i \in \mathcal{F}} x_{ij} = 1, \quad \text{for each } j \in \mathcal{D}, \tag{1.2}$$

$$x_{ij} \leq y_i, \quad \text{for each } i \in \mathcal{F},\ j \in \mathcal{D}, \tag{1.3}$$

$$x_{ij} \geq 0, \quad \text{for each } i \in \mathcal{F},\ j \in \mathcal{D}, \tag{1.4}$$

$$x_{ij} \quad \text{integer}, \quad \text{for each } i \in \mathcal{F},\ j \in \mathcal{D}. \tag{1.5}$$

In effect, we force each variable to be either 0 or 1. We may interpret setting $y_i = 1$ to mean that the facility at location i has been opened, whereas if we set $x_{ij} = 1$, then the client at location j is assigned to a facility at location i. The constraints (1.2) ensure that each client is assigned to some location, whereas the constraints (1.3) ensure that if a client at location j is assigned to a location i, then indeed, a facility has been opened a location i. So, this is an integer programming formulation of the uncapacitated facility location problem. However, the problem is *NP*-hard, and hence no polynomial-time algorithm to solve it will exist unless *P*=*NP*, which is widely viewed as being extremely unlikely.

If we drop the restriction that the variables in (1.1)–(1.4) must take integer values, then we obtain a linear programming (LP) problem. When we drop the integrality restriction in this way, we obtain a *linear programming relaxation* for a problem. Unlike integer programming, there are efficient algorithms known to solve linear programs to optimality. Thus, we can solve the linear programming relaxation of the uncapacitated facility location problem in polynomial time. This leads to one of the most generally applicable methods for designing approximation algorithms: the approach of *LP rounding*. In this approach, we first formulate a problem as an integer program, then consider its linear programming relaxation, solve the relaxation to optimality, and then attempt to round the fractional solution obtained in this way to an integer one (using some polynomial-time procedure). If we can obtain a feasible integer solution that has objective function value within a factor of ρ of the value of the fractional solution that has been rounded, then the resulting algorithm is a ρ-approximation algorithm. The first approximation algorithm for our location problem that we will present will be an LP rounding algorithm.

Unfortunately, LP rounding is not always the method of choice. The main reason for this is that although linear programming relaxations can be solved reasonably efficiently, we are often interested in large inputs, and the LPs that must be solved are quite often huge – too huge to solve within a reasonable amount of time. We can still devise an approximation algorithm based on linear programming, however. For every linear program: minimize $c^T x$ subject to $Ax = b$, $x \geq 0$, there is a dual linear program: maximize $y^T b$ subject to $y^T A \leq c^T$. The strong theorem of linear programming duality states that, provided an optimal solution exists for the original LP, the optimal value of the primal and dual linear programs must be equal. (The reader is referred to the textbook of Chvátal [**2**] for a thorough introduction to linear programming.)

For example, the dual linear program for the linear relaxation (1.1)–(1.4) of the uncapacitated facility location problem is as follows:

$$\text{Maximize} \sum_{j \in \mathcal{D}} v_j \tag{1.6}$$

subject to

$$\sum_{j \in \mathcal{D}} w_{ij} \leq f_i, \quad \text{for each } i \in \mathcal{F}, \tag{1.7}$$

$$v_j - w_{ij} \leq c_{ij}, \quad \text{for each } i \in F,\ j \in \mathcal{D}, \tag{1.8}$$

$$w_{ij} \geq 0 \quad \text{for each } i \in \mathcal{F},\ j \in \mathcal{D}. \tag{1.9}$$

This dual can be motivated in the following way. Suppose that we wish to obtain a lower bound for our input to the uncapacitated facility location problem. If we reset all fixed costs f_i to 0, and solve this input, then clearly we get a (horrible) lower bound: each client $j \in \mathcal{D}$ gets assigned to its closest facility at a cost of $\min_{i \in \mathcal{F}} c_{ij}$. Now suppose we do something a bit less extreme. Each location $i \in \mathcal{F}$ decides on a given cost-sharing of its fixed cost f_i. Each location $j \in \mathcal{D}$ is allocated a share w_{ij} of the fixed cost; if j is assigned to an open facility at i, then it must pay an additional fee of w_{ij} (for a total of $c_{ij} + w_{ij}$), but the explicit fixed cost of i is once again reduced to 0. Of course, we insist that each $w_{ij} \geq 0$, and $\sum_{j\in\mathcal{D}} w_{ij} \leq f_i$ for each $i \in \mathcal{F}$. But this is still an easy input to solve: each $j \in \mathcal{D}$ incurs a cost $v_j = \min_{i\in\mathcal{F}}(c_{ij} + w_{ij})$, and the lower bound is $\sum_{j\in\mathcal{D}} v_j$. Of course, we want to allocate the shares so as to maximize this lower bound, and this maximization problem is precisely the LP dual.

The duality theory of linear programming leads to the so-called *primal-dual method* for designing approximation algorithms. Suppose that we design an algorithm that simultaneously constructs a feasible integer solution to our integer programming formulation, and a feasible solution to the dual of its linear programming relaxation. If the algorithm runs in polynomial time, and we prove that the objective function value of the integer solution is always within a factor of ρ of the (dual) objective function value of the feasible dual solution constructed, then we have obtained a ρ-approximation algorithm. To see this, note that the objective function value of *any* feasible dual solution is a lower bound on the optimal LP value (dual or primal, they are the same), which is in turn a lower bound on the optimal value to our integer program. Hence, if we find a solution of value at most ρ times the value of a feasible dual solution, then it is also at most ρ times the integer optimal value. We shall describe a primal-dual approximation algorithm for the uncapacitated facility location problem in Section 3.

Unfortunately, this is not typically the type of approach employed when a practitioner wishes to compute good solutions fast. A much more common approach is to base the design of an algorithm on the notion of *local search*. Given a feasible solution for a combinatorial optimization problem, it is often natural to think of ways in which one might slightly modify one's current solution to obtain a new solution. For example, in the uncapacitated facility location problem, one might consider opening one extra facility, or closing one of the facilities already open (and then readjusting the assignment of clients to facilities accordingly).

One can view the set of all feasible solutions as a large *neighborhood graph*: each node in the graph corresponds to a feasible solution, and two nodes (i.e., feasible solutions) are adjacent if one can be obtained from the other by one of the specified perturbations. (For simplicity, we shall assume that the perturbations are reversible, and hence the graph is undirected.) A *local optimum*, with respect to some neighborhood structure, is a feasible solution whose cost is no more than all of its neighbors. One naive approach to finding a good solution is to start with an arbitrary feasible solution, and repeatedly check if there is some neighboring solution with better objective function value, until one finds a local optimum. Such an algorithm is called a *local search procedure*. Remarkably, these algorithms often perform very well in practice, especially when enhanced in various ways with additional randomization within the algorithm; simulated annealing is one approach to adding randomization to such algorithms. (The reader is referred to the volume of Aarts and Lenstra [**1**] for an in-depth treatment of this subject.) In Section 4, we shall also present a simple local search algorithm that can be shown to be a 3-approximation algorithm.

The *k-median problem* is closely related to the uncapacitated facility location problem. In this problem, there is a set $\mathcal{N}$ of locations, and the aim is to select k of them (as "medians"), and assign each point in $\mathcal{N}$ to one of the selected medians so as to minimize the total assignment cost. In other words, if we start with the uncapacitated facility location problem, and consider the case in which $\mathcal{D} = \mathcal{F} = \mathcal{N}$, set each $f_i = 0$, but impose the additional constraint that $\sum_{i \in \mathcal{N}} y_i = k$, then we obtain the k-median problem. For any input to the k-median problem, we can view it as an input to the uncapacitated facility location problem, except that there are no facility costs given. Suppose we set the facility costs equal to some uniform value z; if $z = 0$, then a reasonable solution to the uncapacitated facility location problem would tend to open a facility at each location; on the other hand, if z is set *very* large, then only one facility would be opened. By adjusting the value of this *Lagrangean multiplier*, it seems natural that we can find a solution that opens k facilities, and hence can serve as a good solution for the k-median problem. In fact, this approach does work, and we will describe this in Section 5.

2. An LP rounding algorithm

We shall first describe a simple procedure to take an optimal solution (x^*, y^*) to the linear programming relaxation for the uncapacitated facility location problem and produce a feasible integer solution $(\bar{x}, \bar{y})$; furthermore, it will be straightforward to prove that the cost of the integer solution is no more than 4 times the cost of the optimal fractional solution. Let (v^*, w^*) denote an optimal solution to the dual linear program.

The algorithm works as follows. First represent the fractional solution (x^*, y^*) as a bipartite graph $G = (\mathcal{F}, \mathcal{D}, E)$ where, for each (i, j) such that $x^*_{ij} > 0$, we include the edge (i, j) in E, i.e., there are two set of nodes $\mathcal{F}$ and $\mathcal{D}$, and each edge in E has one endpoint in each set. We shall rely on an important property of optimal solutions of linear programs, known as complementary slackness: this states that whenever a variable is positive in an optimal solution, then for any optimal solution to the dual, the constraint in the dual that corresponds to this variable must be satisfied with equality. In particular, this means that for each edge $(i, j) \in E$, we have that $v^*_j - w^*_{ij} = c_{ij}$. Since $w^*_{ij} \geq 0$, it follows that $c_{ij} \leq v^*_j$; for each edge in G, we know that the associated assignment cost is not too large.

The algorithm partitions the graph G into clusters, and then, for each cluster, opens one facility. The clusters are constructed iteratively as follows. Among all clients that have not already been assigned to a cluster, let j' be the client j for which v^*_j is smallest. This cluster consists of j', all neighbors of j' in G, and all of their neighbors as well (that is, all nodes j such that there exists some i for which (i, j) and (i, j') are both in E). Within this cluster, we open the cheapest facility i'. After all of the clusters have been constructed, each client is then assigned to its nearest opened facility.

We first show that each client has a nearby facility that has been opened. Consider a client j, let i' denote the facility opened in the cluster that contains j, and let j' be the node with minimum v^*-value used in forming the cluster (see Figure 1). In other words, we know that there exists a path in G from j to i' consisting of an edge connecting i' and j' (of cost at most $v^*_{j'}$), an edge connecting j' and some node i (of cost at most $v^*_{j'}$), and an edge connecting i and j (of cost at most v^*_j). Hence, by the triangle inequality, the cost of assigning j to i' is at most $2v^*_{j'} + v^*_j$. Since j' was chosen as the remaining client with minimum v^*-value, it follows that $v^*_{j'} \leq v^*_j$, and so the cost of assigning j to i' is at most $3v^*_j$. Hence, the total assignment cost of the solution found is at most $3 \sum_{j \in \mathcal{D}} v^*_j$, which is 3 times the optimal objective value of the (dual) linear program.

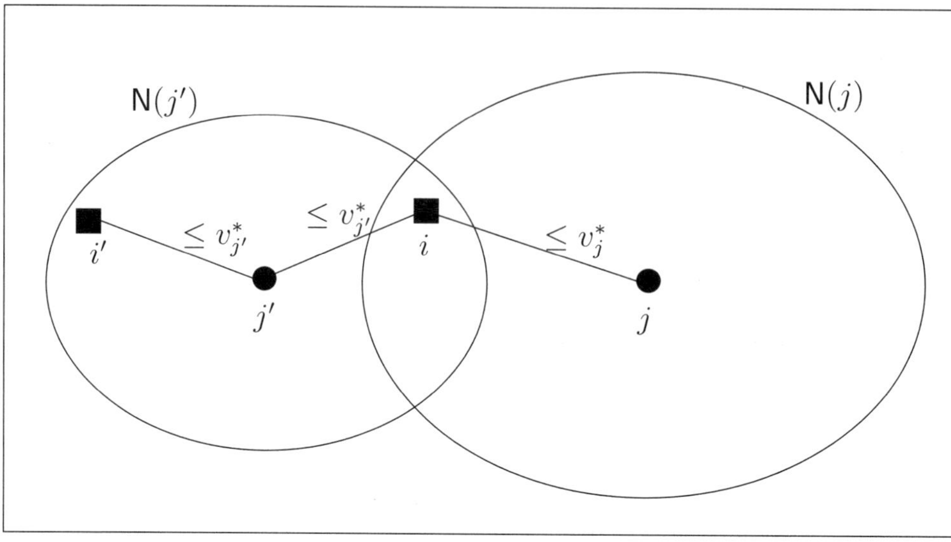

FIGURE 1. Bounding the assignment cost of j; $\mathsf{N}(j)$ and $\mathsf{N}(j')$ denote the set of neighbors of j and j', respectively, in G.

Next we bound the total cost of the facilities that are opened by the algorithm. Consider the first cluster formed, and let j' be the node with minimum v^*-value used in forming it. We know that $\sum_{i:(i,j')\in E} x^*_{ij'} = 1$. Since the minimum of a set of values is never more than a weighted average of them, the cost of the facility selected is

$$f_{i'} \leq \sum_{i:(i,j')\in E} x^*_{ij'} f_i \leq \sum_{i:(i,j')\in E} y^*_i f_i,$$

where the last inequality follows from constraint (1.3). That is, for the first cluster, the cost of the facility opened in that cluster is no more than the fractional cost incurred by all of the facilities (fractionally) opened in the cluster in the LP optimum.

Observe that, throughout the execution of the algorithm, each location $j \in \mathcal{D}$ that has not yet been assigned to some cluster has the property that each of its neighbors i must also remain unassigned. Hence, for *each* cluster, the cost of its open facility is at most the cost that the fractional solution assigned to nodes in $\mathcal{F}$ within that cluster. Hence, in total,

$$\sum_{i\in\mathcal{F}} f_i \bar{y}_i \leq \sum_{i\in\mathcal{F}} f_i y^*_i;$$

that is, the total facility cost of the rounded solution is at most the facility cost incurred by the optimal fractional solution (which is, of course, at most the total cost of the optimal LP solution). Thus, the total cost of the solution found by the rounding algorithm is at most 4 times the cost of the optimal solution to the LP.

There is a simple way to improve the algorithm, which relies on a technique known as *randomized rounding*. Consider a cluster formed by first selecting the client j'. In the previous algorithm, we opened the cheapest neighbor of j'. Instead, choose a neighbor i of j' with probability $x^*_{ij'}$, and open that facility. Furthermore, for each facility i in the cluster, open an additional facility at i with probability $y^*_i - x^*_{ij'}$. Finally, for each facility i that is not in any cluster, open that facility with probability y^*_i. Once again, assign each client to its nearest open facility. This is a randomized algorithm; the algorithm "tosses coins" and the output of the algorithm is a random variable. It is easy to see that the expected facility cost incurred by this solution is at most the facility cost incurred by the optimal LP solution. In this case, one can prove a tighter bound on the expected assignment cost incurred (since, for each client j, with probability at least $1 - 1/e$, one of the neighbors of j in G is an open facility). Overall, one can prove that the expected cost of the solution found by this randomized algorithm is at most $1+3/e$ times the cost of the optimal LP solution. Furthermore, one can apply a simple "greedy rule" to obtain, without randomization, a solution of cost no greater than this expected value, and hence obtain a $(1+3/e)$-approximation algorithm. (This derandomization method is generally called the *method of conditional expectations.*)

3. A primal-dual algorithm

Primal-dual algorithms have the advantage that one need not solve a linear program, while still exploiting the structure that a linear programming relaxation brings to many problems. As described above, we will construct a feasible integer solution for the facility location problem, and at the same time construct a feasible solution to the dual of the linear programming relaxation. To prove the performance guarantee of the algorithm, we will bound the cost of the (primal) solution computed by a constant factor of the objective function of the dual solution computed.

We shall describe an algorithm that works in two phases. In the first phase, we compute a feasible dual solution. This dual solution will be used to construct a graph, much as the optimal LP solution was used in the previous section. In the second phase, the graph will be decomposed into clusters (though in a manner different from the method used above), where again we open one facility in each cluster. Finally, each client j will be assigned to the nearest opened facility.

The procedure to compute a feasible dual solution is what is often called a *dual ascent method.* We start with a feasible solution, where all of the dual variables are set to 0, and then increase the value assigned to some of the variables while maintaining a feasible solution. In particular, there is a continuous notion of time that evolves throughout the execution of the algorithm. At time $t = 0$, all of the dual variables are set to 0. As time progresses, each variable v_j will continue to be equal to the time t, until some point in the algorithm, when that variable is frozen, and that variable remains fixed for the remainder of the algorithm. For each client j, the final value of its dual variable, which we shall denote $\tilde{v}_j$, denotes the time at which that variable is frozen. Initially, we shall keep all of the variables w_{ij} set to 0.

As time progresses, there will be three types of events that determine the course of the algorithm. Suppose that we have reached a time t for which, for some client, $v_j = t$ is equal to c_{ij}, for some facility location $i \in \mathcal{F}$. We can view this as a physical process, where the sphere of influence for client j has reached the facility i. Note that if we allow the algorithm to proceed unchanged, then at subsequent times, we shall set v_j to a value for which constraint (1.8) is violated, and we will no longer maintain the invariant that the current dual solution is feasible. Instead, we also increase w_{ij} at a rate of 1 unit per unit of time elapsed; that is, we maintain that v_j is equal to the current time t, and also maintain that $c_{ij} + w_{ij} = v_j$. If one views v_j as being the composite cost of serving the client at location j, once the current budget for j is sufficient to pay to connect to a facility at location i, then the excess can go to the share of the fixed cost for i that location j is willing to pay.

The second type of event occurs when, for some facility location i, the total of the shares being offered to it by the clients is sufficient to pay the fixed cost f_i. Stated more algebraically, we have increased the values w_{ij} so that the constraint (1.7) now holds with equality. Clearly, when this happens, we can no longer increase each w_{ij} for a client j for which $v_j \geq c_{ij}$. For each such client j, we freeze all dual variables v_j and $w_{i'j}$ (for all i' including i) associated with the client j (at their current values). We shall say that facility i is now *paid for*, and we let t_i denote the time at which facility i is paid for.

In the final type of event, the dual variable v_j for a client j increases to be equal to c_{ij}, where the facility i is already paid for (by the shares of *other* clients). Unlike in the first type of event, we cannot increase w_{ij} anymore, and hence we freeze all dual variables v_j and $w_{i'j}$ at their current values. Although we have described this procedure in terms of a continuous notion of time, it is straightforward to view it in terms of the events that occur, by calculating the next (discrete) event (and advancing the time to that moment).

We continue this dual ascent process until all dual variables v_j have been frozen. Let $(\tilde{v}, \tilde{w})$ denote the final solution computed by this procedure. Construct the graph $G = (\mathcal{F}, \mathcal{D}, E)$ where, for each (i, j) such that $\tilde{w}_{ij} > 0$, we include the edge (i, j) in E. Once again, we have a similar property as in the LP algorithm: for each edge $(i, j) \in E$, we have that $c_{ij} + \tilde{w}_{ij} = \tilde{v}_j$. Since $\tilde{w}_{ij} > 0$, it follows that $c_{ij} < \tilde{v}_j$; for each edge in G, we know that the associated assignment cost is not too large.

Once again, the algorithm partitions the graph G into clusters, and then, for each cluster, opens one facility. The clusters are constructed iteratively as follows. Among all facilities that are paid for and that have not yet been assigned to a cluster, let i' be the facility that is paid for earliest; that is, t_i is smallest. The new cluster consists of i', all neighbors of i' in G, and all of their neighbors as well (that is, all facility locations i such that there exists some j for which (i, j) and (i', j) are both in E (or in other words, are such that both $\tilde{w}_{ij} > 0$ and $\tilde{w}_{i'j} > 0$). We open the facility i'. This continues until all paid-for facilities have been assigned to a cluster. After all of the clusters have been constructed, each client is then assigned to its nearest opened facility.

We now analyze this algorithm. We will prove that the total cost of the solution found is at most $3 \sum_{j \in \mathcal{D}} \tilde{v}_j$, and hence is within a factor of three of optimal. In fact, we will prove a stronger property: we will show that not only is the total cost at most this much, if we consider 3 times the facility cost incurred plus the total assignment cost, then even this quantity is no more than $3 \sum_{j \in \mathcal{D}} \tilde{v}_j$. (We will explain in Section 5 why this is useful.)

Consider any client j that is adjacent in G to a facility i that has actually been opened. Note that our algorithm has ensured that, for any client j, there is at most one such facility. The assignment cost for this client is therefore at most c_{ij}. But by dual feasibility, we see that even $3(c_{ij} + \tilde{w}_{ij}) = 3\tilde{v}_j$. We shall charge each such client only $c_{ij} + 3\tilde{w}_{ij}$, which is at most $3\tilde{v}_j$. This charge pays for its assignment cost, and 3 times its share of the fixed cost f_i.

On the other hand, consider the open facility i; each of its neighbors j' in G will pay for three times $\tilde{w}_{ij'}$, its share of the fixed cost f_i. But these neighbors are the only clients that contribute *any* positive share to the fixed cost of location i in the dual solution $(\tilde{v}, \tilde{w})$, and since i was opened, the facility at location i must have been paid for. Hence, the total charge incurred by its neighbors must pay for all of their assignment costs, as well as $3f_i$.

Consider any client that is not adjacent to an open facility. For each such client j, we will argue that its assignment cost is most $3\tilde{v}_j$, and this is sufficient to complete the proof of the performance guarantee. Since the dual variable v_j was frozen at some point by the algorithm, there must be some paid-for facility i for which $\tilde{v}_j \geq c_{ij}$. If i has been opened, then the assignment cost incurred for j is clearly at most $c_{ij} \leq \tilde{v}_j \leq 3\tilde{v}_j$.

Suppose that i has not been opened. Then it must belong to the cluster of some other open facility i'; furthermore, there is a client j' that is contributing a positive share to the facility costs of both i and i'. Consider the moments in time that facilities i and i' have been paid for, denoted t_i and $t_{i'}$, respectively. Since facility i is in the cluster started by facility i', we have that $t_{i'} \leq t_i$. Furthermore, since client j' contributes a positive share to the fixed cost of both i and i', we have that $c_{ij'} < t_i$ and $c_{i'j'} < t_{i'}$. Furthermore, facility i must be paid for no later than the time that client j is frozen; hence, $t_i \leq \tilde{v}_j$. Hence, the distance from client j to the open facility i' is at most

$$c_{ij} + c_{ij'} + c_{i'j'} < \tilde{v}_j + t_i + t_{i'} \leq \tilde{v}_j + 2t_i \leq 3\tilde{v}_j.$$

Hence, we have shown that this primal-dual algorithm is a 3-approximation algorithm.

4. A local search algorithm

Local search is one of the most effective techniques for designing heuristic procedures to compute near-optimal solutions. Unfortunately, in most cases we cannot prove a strong guarantee on the quality of the solution found. LP rounding algorithms and primal-dual algorithms compute both a feasible integer solution and a lower bound; hence, even without an *a priori* guarantee, these approaches generate, on an input by input basis, a means to provide an *a postiori* bound on the performance of the algorithm. Unfortunately, no comparable information is generated by a local search procedure. Thus, it is even more important to be able to prove a performance guarantee for a local search algorithm that holds for all inputs.

In describing a feasible solution to the uncapacitated facility location problem, it is natural to focus on the set of facilities that have been opened, since we can assume that each client is assigned to its nearest open facility. Thus, our aim is to find a set S^* for which the total cost is minimum. However, in describing the local search procedure, we will maintain both a set of open facilities, and a particular assignment of clients to open facilities.

The algorithm starts by choosing one facility location $i_0 \in \mathcal{F}$ arbitrarily, and then setting $S = \{i_0\}$ (and each client is assigned to that facility). In each iteration, one can change S in one of two possible ways: (1) an *add move*, in which some facility location $i \in \mathcal{F}$ is added to S; and (2) a *swap move*, in which we first pick one facility location i^* to be added to S, and then for each $i \in S$, we see whether there is an improving net effect if we delete i and reassign to i^* *all* of the clients that i currently serves; if so, we delete i and make this reassignment. It is easy to check whether there is a move of either type that yields an improved solution; if so, the algorithm modifies S by (an arbitrarily chosen) one of these allowed moves. We shall call S a *local optimum* if there is no move of either type that yields an improved solution.

Let S^* denote an optimal solution, and let F^* and C^* denote the total facility and assignment costs associated with it. Consider an arbitrary feasible solution S, and let F and C denote its associated facility and assignment costs. We will prove that the following two claims hold:

(1) if no add move from S decreases the cost, then $C \leq C^* + F^*$.
(2) if no swap move from S decreases the cost, then $F \leq F^* + C^* + C$.

We first argue that if we establish both of these claims, then this will imply a performance guarantee on the objective function value of an arbitrary local optimum S. There is no improving add move from S, and so, $C \leq C^* + F^*$. Similarly, there is no improving swap move from S, and hence $F \leq F^* + C^* + C$. If we add twice the former inequality to the latter one, we see that $2C + F \leq 3F^* + 3C^* + C$, and hence $C + F \leq 3F^* + 3C^*$. But this means that the cost of the local optimum S is at most three times the optimal cost.

So now we must prove that the two claims hold. Before proceeding with their proofs, we first introduce a bit more notation. For the optimal solution S^*, let $\iota^*(j)$ denote the facility $i \in S^*$ to which j is assigned, for each $j \in \mathcal{D}$. Furthermore, for each $i \in S^*$, let D_i^* denote the set of locations j for which $\iota^*(j) = i$. Similarly, for the current solution S, let $\iota(j)$ denote the facility $i \in S$ to which j is currently assigned, for each $j \in \mathcal{D}$, and let D_i denote the set of locations $j \in \mathcal{D}$ currently assigned to i (i.e., $\iota(j) = i$).

The proof of the first claim, about the add move, is relatively straightforward. Suppose that no add move exists that improves the cost of the current solution S. For each $i \in S^*$, consider the net change in cost of adding i to the current solution. Ordinarily, one would assume that not only is i added to the solution, but each client j that is closer to i than to its current facility is reassigned to i. We will consider instead an inferior move, in which i is added to the solution, and each location $j \in D_i^*$ is assigned to i, whether or not this is an improvement (and these are the only changes in the assignment). Of course, none of these moves will improve the overall cost of the solution either. That is, we have that

$$0 \leq f_i + \sum_{j \in D_i^*} c_{ij} - c_{\iota(j)j}. \tag{4.1}$$

By adding (4.1) for each $i \in S^*$, we get that

$$\begin{aligned} 0 &\leq \sum_{i \in S^*} f_i + \sum_{j \in D_i^*} (c_{ij} - c_{\iota(j)j}) \\ &= \sum_{i \in S^*} f_i + \sum_{i \in S^*} \sum_{j \in D_i^*} c_{ij} - \sum_{i \in S^*} \sum_{j \in D_i^*} c_{\iota(j)j} \\ &= F^* + C^* - C, \end{aligned}$$

where the last equality follows from the fact that D_i^*, $i \in S^*$, constitutes a partition of $\mathcal{D}$ (based on the facility to which the clients are assigned in the optimal solution). Of course, this implies that $C \leq F^* + C^*$, which is what we wished to prove.

The proof of the second claim is somewhat more complicated, but has the same flavor; rather than analyzing the original swap move, one analyzes an inferior (and unimplementable) modified swap move. The modified swap move can be described as follows. For each $i \in S$, consider the set $\hat{S} = \cup_{j \in D_i} \iota^*(j)$, and find the facility $i^* \in \hat{S}$ such that c_{ii^*} is smallest. In this case, we say that i *maps to* i^*. For each $i^* \in S^*$, let S_{i^*} denote the set of locations $i \in S$ that map to i^*. In our *modified swap move*, we add i^*, delete S_{i^*}, and assign each client $j \in \cup_{i \in S_{i^*}} D_i$ to i^*.

For the modified swap move, we are closing the facilities at S_{i^*} and instead opening a facility at i^*. Therefore, we need to reassign all clients that were currently assigned to a facility $i \in S_{i^*}$. The natural thing is to reassign them to the newly opened facility at i^*, and we want to upper bound the increase in their assignment cost by this change. We shall argue that, for each $j \in D_i$,

$$c_{i^*j} \leq 2c_{ij} + c_{\iota^*(j)j}; \tag{4.2}$$

in other words, the reassignment changes the assignment cost of j from c_{ij} to at most $c_{ij} + (c_{ij} + c_{\iota^*(j)j})$, or equivalently, the increase is at most $c_{\iota(j)j} + c_{\iota^*(j)j}$ (since $\iota(j) = i$). The cost from i^* to j can be upper bounded by the sum of the cost from i^* to i, and the cost from i to j; that is,

$$\begin{aligned} c_{i^*j} &\leq c_{i^*i} + c_{ij} \\ &\leq c_{\iota^*(j)i} + c_{ij} \\ &\leq c_{\iota^*(j)j} + c_{ij} + c_{ij}, \end{aligned}$$

where first and last inequalities follow from the triangle inequality, and the middle one follows from our choice of i^*.

This modified swap move has the limitation that, having fixed i^* as the new facility, and the one to which all reassignments will be made, we simply delete all current facilities i that map to i^* rather than focusing on the set of current facilities for which the savings in deleting i outweighs the additional assignment cost. Consequently, if there is no improving swap move, there is also no improving modified swap move.

Consider the change in cost associated with a modified swap move associated with $i^* \in S^*$. This change is

$$f_{i^*} - \sum_{i \in S_{i^*}} f_i + \sum_{i \in S_{i^*}} \sum_{j \in D_i} (c_{\iota(j)j} + c_{\iota^*(j)j}) \geq 0,$$

where the inequality follows from that fact that the move is not an improving one. By summing this inequality over all choices $i^* \in S^*$, we get that

$$\begin{aligned} 0 &\leq \sum_{i^* \in S^*} \left[f_{i^*} - \sum_{i \in S_{i^*}} f_i + \sum_{i \in S_{i^*}} \sum_{j \in D_i} (c_{\iota(j)j} + c_{\iota^*(j)j}) \right] \\ &= \sum_{i \in S^*} f_i - \sum_{i \in S} f_i + \sum_{j \in \mathcal{D}} (c_{\iota(j)j} + c_{\iota^*(j)j}) \\ &= F^* - F + C + C^*, \end{aligned}$$

which implies the second claim.

We have shown that any locally optimal solution (with respect to these two types of moves) has cost at most three times the optimum. We have not, however, proved that this local search algorithm is a 3-approximation algorithm; we would also need to prove a polynomial upper bound on the number of moves needed to reach a local optimum. We do not know if such a bound holds, and hence we settle for a weaker guarantee. Instead of focusing on whether there is a move that improves the solution at all, we can instead focus on whether there is a move that improves the solution by a $1+\epsilon$ factor, for some constant $\epsilon > 0$. Now it is easy to show that a polynomial number of $(1+\epsilon)$ improving moves suffices to reach such a "nearly" local optimal solution (where there might be improving moves, but none that improves the cost by a $1+\epsilon$ factor). Furthermore, an analysis similar to the one above shows that this approach leads to a $(3+\epsilon)$-approximation algorithm, for any constant $\epsilon > 0$.

5. Lagrangean relaxation and the k-median problem

Lagrangean relaxation is a technique often applied in combinatorial optimization, where the constraints of a problem can be partitioned into two parts: the "easy" ones and the "hard" ones. If one ignores the hard constraints, then the optimization problem (subject to just the easy constraints) becomes (more) easily solvable. Thus, one can add a Lagrangean multiplier (or dual variable) for each hard constraint that penalizes the violation of that constraint, and this Lagrangean penalty function can then be added to the original objective function; we then optimize the combined problem to find the "appropriate" value for the Lagrangean multiplier.

Consider the natural integer programming formulation of the k-median problem:

$$\text{Minimize } \sum_{i\in\mathcal{N}}\sum_{j\in\mathcal{N}} c_{ij}x_{ij} \tag{5.1}$$

subject to

$$\sum_{i\in\mathcal{N}} x_{ij} = 1, \quad \text{for each } j\in\mathcal{N}, \tag{5.2}$$

$$\sum_{i\in\mathcal{N}} y_i = k, \tag{5.3}$$

$$x_{ij} \le y_i, \quad \text{for each } i\in\mathcal{N},\ j\in\mathcal{N}, \tag{5.4}$$

$$x_{ij} \ge 0, \quad \text{for each } i\in\mathcal{N},\ j\in\mathcal{N}. \tag{5.5}$$

If we relax the constraint (5.3), and introduce a penalty z for violating the constraint, thereby adding the term $z(\sum_{i\in\mathcal{N}} y_i - k)$, we obtain the uncapacitated facility location problem in which the fixed costs for opening the facilities are now equal to the common value z (ignoring the $-kz$ term in the objective function, which is a constant for any fixed value of z).

Given an input to the k-median problem, we can use this connection to the uncapacitated facility location problem in the following way. Suppose that we apply the primal-dual algorithm from Section 3 to our input with a small common fixed z. Then the algorithm is likely to open many (if not all) of the facilities. On the other hand, if the value z is sufficiently large, then the algorithm will open only one facility. Suppose that we have set z so that the algorithm outputs a solution with exactly k open facilities. Let $(\tilde{x},\tilde{y})$ and $(\tilde{v},\tilde{w})$ denote the primal and dual solutions constructed for this value $\tilde{z}$. If we let z denote the variable dual to the additional constraint (5.3), then the dual of the linear programming relaxation given above is:

$$\text{Maximize} \sum_{j\in\mathcal{N}} v_j \quad - \quad kz$$

subject to

$$\begin{aligned} \sum_{j\in\mathcal{N}} w_{ij} &\le z, && \text{for each } i \in \mathcal{N}, \\ v_j - w_{ij} &\le c_{ij}, && \text{for each } i \in F,\ j \in \mathcal{N}, \\ w_{ij} &\ge 0 && \text{for each } i \in \mathcal{N},\ j \in \mathcal{N}. \end{aligned}$$

Since $(\tilde{x}, \tilde{y})$ opens exactly k facilities, this is a feasible integer solution for the k-median LP formulation. Furthermore, since we have used the variable z to reflect the common fixed cost of opening a facility, $(\tilde{v}, \tilde{w}, \tilde{z})$ is a feasible solution for the dual LP. The performance guarantee that we proved for the primal-dual algorithm for the uncapacitated facility location problem implies that

$$3\sum_{i\in\mathcal{N}} \tilde{z}\tilde{y}_i + \sum_{i\in\mathcal{N}}\sum_{j\in\mathcal{N}} c_{ij}\tilde{x}_{ij} \le 3\sum_{j\in\mathcal{N}} \tilde{v}_j,$$

or equivalently, that

$$\sum_{i\in\mathcal{N}}\sum_{j\in\mathcal{N}} c_{ij}\tilde{x}_{ij} \le 3\sum_{j\in\mathcal{N}} \tilde{v}_j - 3\tilde{z}(\sum_{i\in\mathcal{N}} \tilde{y}_i) = 3(\sum_{j\in\mathcal{N}} \tilde{v}_j - kz).$$

This is, we have found a feasible integer solution and a feasible dual solution for the k-median problem that are within a factor of three of each other.

By focusing on one problem domain, and exploring a few algorithmic techniques in depth, we have attempted to give the reader a sense of the mathematics involved in this area of algorithm design for discrete optimization problems. We have focused on results that were easy to present, rather than the state of the art for each of the approaches. Many more results are summarized in [**5**], but even that recent survey misses several interesting results from the past couple of years.

References

[1] E. L. Aarts and J. K. Lenstra, eds. *Local search in combinatorial optimization*, Princeton University Press, Princeton, 2003.

[2] V. Chvátal, *Linear programming*, W.H. Freeman, NY, 1983.

[3] D. S. Hochbaum, ed., *Approximation algorithms for NP-hard problems*, PWS Publishing Company, Boston, MA, 1997.

[4] D. B. Shmoys, *Computing near-optimal solutions to combinatorial optimization problems*, in: W. Cook, L. Lovász, and P. Seymour, eds., *Combinatorial optimization*, AMS, Providence, RI, 1994, pages 355–397.

[5] D. B. Shmoys, *Approximation algorithms for facility location problems*, in: K. Jansen, S. Khuller, eds., *Approximation Algorithms for Combinatorial Optimization, Third International Workshop, APPROX 2000, Lecture Notes in Computer Science 1913*, Springer, Berlin, 2000, pages 27–33.

[6] V. V. Vazirani, *Approximation algorithms*, Springer Verlag, Berlin, 2001.

School of Operations Research & Industrial Engineering and Department of Computer Science, Cornell University, Ithaca, New York 14853

Current address: School of Operations Research & Industrial Engineering and Department of Computer Science, Cornell University, Ithaca, New York 14853

E-mail address: `shmoys@cs.cornell.edu`

Proceedings of Symposia in Applied Mathematics
Volume **61**, 2004

Algebraic Recipes for Integer Programming

Bernd Sturmfels

ABSTRACT. Integer programming is concerned with solving linear systems of equations over the non-negative integers. The basic question is to find a solution which minimizes a given linear objective function for a fixed right hand side. Here we also consider parametric versions where the objective function and the right hand side are allowed to vary. The main emphasis is on Gröbner bases, rational generating functions, and how to use existing software packages. Concrete applications to problems in statistical modeling will be presented.

1. An Introductory Coin Problem

This lecture is about solving linear equations over the non-negative integers. Our point of departure is the integer programming problem in *standard form*:

$$\text{Minimize } c \cdot u \text{ subject to } A \cdot u = b \text{ and } u \in \mathbb{N}^n. \tag{1.1}$$

The given instance consists of an integer matrix $A \in \mathbb{Z}^{d \times n}$, a row vector $c \in \mathbb{Z}^n$ and a column vector $b \in \mathbb{Z}^d$. The unknown is the column vector $u = (u_1, \ldots, u_n)$. What makes the problem hard is the requirement that the u_i be non-negative integers.

As an example consider the following simple *coin problem*. Suppose you are carrying a large collection of coins in your pocket. The allowed coins are pennies (1 cent), nickels (5 cents), dimes (10 cents) and quarters (25 cents). The problem is to replace your "portfolio" by an equal number of coins having the same monetary value, but such that the number of nickels plus the number of quarters is minimized.

This problem can be expressed in the standard form (1.1) by setting

$$d = 2,\, n = 4, \qquad A = \begin{pmatrix} 1 & 1 & 1 & 1 \\ 1 & 5 & 10 & 25 \end{pmatrix} \text{ and } c = (0, 1, 0, 1). \tag{1.2}$$

The right hand side vector $b = \binom{b_1}{b_2}$ is left unspecified. Its coordinates b_1 and b_2 are the number of coins and value (in pennies) of the portfolio respectively.

EXAMPLE 1.1. For instance, if $b = \binom{10}{114}$, then we seek to express one dollar and fourteen cents with ten coins. The optimal solution to this instance of (1.1) is $u = (4, 2, 0, 4)$ and the *optimal value* is $c \cdot u = 6$. In words, you can take four

---

1991 *Mathematics Subject Classification.* Primary 90C10; Secondary 13P10, 52B20, 62H17.
This work was done while the author held a Hewlett Packard Visiting Research Professorship 2003/2004 at MSRI Berkeley. He was also partially supported by NSF Grant DMS-0200729.

pennies, two nickels and four quarters to make one dollar and fourteen cents with ten coins, but it is impossible to do it with less than six nickels or quarters. □

A parametric solution to our problem is provided by the *Gröbner basis*

$$(1.3) \qquad \mathcal{G} \quad = \quad \{\, n^3q - d^4,\ n^6 - p^5q,\ n^3d^4 - p^5q^2,\ p^5q^3 - d^8 \,\}.$$

Our Gröbner basis in (1.3) is expressed as a set of *monomial differences*, which is how they usually appear in computer algebra systems. We note that there are two alternative but entirely equivalent ways of writing our Gröbner basis. In the optimization literature, it is more common to express $\mathcal{G}$ as a *set of lattice vectors*;

$$(1.4) \qquad \mathcal{G}' \quad = \quad \{\ (0,3,-4,1),\ (-5,6,0,-1),\ (-5,3,4,-2),\ (5,0,-8,3)\ \}.$$

Here each lattice vector is regarded as the difference of two non-negative vectors with disjoint support. This mirrors the monomial differences in (1.3). For instance,

$$(0,3,-4,1) \;=\; (0,3,0,1) - (0,0,4,0) \qquad \longleftrightarrow \qquad n^3q - d^4.$$

The Gröbner basis is a set of *exchange rules* which you can use to successively improve your portfolio. For instance, the first rule $n^3q - d^4$ says that you can replace three nickels and one quarter with four dimes. Each of the four moves in $\mathcal{G}$ changes neither the number of coins nor their value but it decreases the objective function value. The crucial property of being a Gröbner basis says that *if none of the exchange rules can be applied then your portfolio is guaranteed to be optimal.*

There is a third way of encoding the Gröbner basis, which will be of importance in Section 4. Namely, we can also express $\mathcal{G}$ as the following *generating function*:

$$(1.5) \qquad \mathcal{G}'' \quad = \quad n^3D^4q + P^5n^6Q + P^5n^3d^4Q^2 + p^5D^8q^3.$$

In the last representation there are two variables for each column of the matrix A, and each monomial represents one exchange rule. The lower case variable represents the gain and the upper case variable represents the loss in the exchange of coins.

In Section 2 we explain how the Gröbner basis is constructed for an arbitrary matrix A and cost function c. Note that the Gröbner basis is independent of the right hand side b, so it provides a parametric solution to the family of integer programs (1.1) for all possible values of b. In Section 3 we discuss the relationship to other notions of test sets in integer programming, including *Hilbert bases* and *Graver bases*, and we introduce Hemmecke's easy-to-use software `4ti2` for computing these test sets. In Section 4 we address complexity issues. In particular, we show how the Gröbner basis can computed in polynomial time when d and n are fixed.

The power of algebraic methods in integer programming stems from the fact that they can answer parametric questions like: *What are all the optimal portfolios in our coin problem ?* Each portfolio is given as a vector $u = (u_1, u_2, u_3, u_4)$ or as a monomial $p^{u_1}n^{u_2}d^{u_3}q^{u_4}$, and we wish to encode all portfolios that are optimal solutions of (1.1) with $b = Au$. The following three conditions are equivalent:

(a) The vector $u \in \mathbb{N}^4$ is an optimal portfolio.
(b) None of the four monomials n^3q, n^6, n^3d^4 or p^5q^3 divides $p^{u_1}n^{u_2}d^{u_3}q^{u_4}$.
(c) $(u_2 \le 2$ or $u_4 = 0)$ and $(u_2 \le 5)$ and $(u_2 \le 2$ or $u_3 \le 3)$ and $(u_1 \le 4$ or $u_4 \le 2)$.

The *Hilbert series of all optimal solutions* is the formal sum of these monomials:

$$\sum_{u \text{ optimal}} p^{u_1}n^{u_2}d^{u_3}q^{u_4}$$

This generating function is equal to the following rational function:

$$\frac{n^3q^3p^5 - n^6qd^4 + n^6d^4 + n^3qd^4 - q^3p^5 + n^6q - n^3d^4 - n^6 - n^3q + 1}{(1-p)(1-n)(1-d)(1-q)}.$$

In Section 4, we will see that such Hilbert series can be computed in polynomial time (for fixed d and n). In Sections 5 and 6 we will focus on applications of integer programming to *statistics*, and we will argue that Gröbner bases and generating functions are useful tools for the optimization problems arising in this context.

The use of Gröbner bases as a tool for integer programming first appeared in the paper [**CT**] by Conti and Traverso. Their approach was further developed in two doctoral dissertations in the Cornell Operations Research Department, written by Thomas (see [**Tho**]) and Hoşten (see [**HS1**]) in 1994 and 1997 respectively. Subsequently, Hoşten and Thomas [**HT2**] developed an algebraic theory of group relaxations, extending the foundational work in integer programming theory which was done by Gomory in the 1960's. These and many other important topics will not be discussed in this lecture, which aims to be introductory and self-contained. Readers wishing to learn more about commutative algebra methods in integer programming are referred to the book [**Stu**] and the survey articles [**HT1**] and [**Tho2**].

2. Gröbner Bases

We consider the integer programming problem in standard form (1.1) where A and c are fixed and b is arbitrary. In this section we further assume that c is *generic* in the sense that (1.1) has a unique optimal solution for every feasible right hand side b. In practise, this can always be accomplished by lexicographically perturbing the given cost vector c. Consider the infinite set of all optimal solutions,

$$\mathrm{Opt}_{A,c} \quad = \quad \big\{ u \in \mathbb{N}^n \,:\, u \text{ is the optimal solution of (1.1) for } b = Au \big\}.$$

Suppose that u and u' are vectors in $\mathbb{N}^n$ such that $u' \leq u$ (coordinatewise) and $u \in \mathrm{Opt}_{A,c}$. Then it can be seen that $u' \in \mathrm{Opt}_{A,c}$. We paraphrase this observation in the following lemma, using the language of partially ordered sets (= posets).

LEMMA 2.1. *The set* $\mathrm{Opt}_{A,c}$ *is an* order ideal *in the partially ordered set* $\mathbb{N}^n$.

A basic result about order ideals in the poset $\mathbb{N}^n$, known as *Dickson's Lemma*, states that the set of minimal elements in the complementary set $\mathrm{Opt}_{A,c}\backslash\mathbb{N}^n$ is finite. We write $\mathrm{Min}(\mathbb{N}^n\backslash\mathrm{Opt}_{A,c})$ for this finite set. Its elements are called the *minimally non-optimal points* of the integer programming family (1.1). Recall that our introductory coin example had precisely four minimally non-optimal points:

$$\mathrm{Min}(\mathbb{N}^4\backslash\mathrm{Opt}_{A,c}) \quad = \quad \big\{ (0,3,0,1),\ (0,6,0,0),\ (0,3,4,0),\ (5,0,0,3) \big\}.$$

For every $g^+ \in \mathrm{Min}(\mathbb{N}^n\backslash\mathrm{Opt}_{A,c})$ there exists a unique vector $g^- \in \mathrm{Opt}_{A,c}$ such that $Ag^+ = Ag^-$. Namely, g^- is the optimal solution to (1.1) with $b = Ag^-$.

DEFINITION 2.2. The *Gröbner basis* for the matrix A and cost vector c is

$$\mathcal{G}_{A,c} \quad = \quad \big\{ \, g^+ - g^- \,:\, g^+ \in \mathrm{Min}(\mathbb{N}^n\backslash\mathrm{Opt}_{A,c}) \, \big\}.$$

This is a finite set of lattice vectors in the kernel of A. We can also regard them as monomial differences in n unknowns $x_1, \ldots, x_n$, or as monomials in $2n$ unknowns $x_1, \ldots, x_n, y_1, \ldots, y_n$, via the encodings

$$(2,13,0,-8,5,-7) \quad \longleftrightarrow \quad x_1^2x_2^{13}x_5^5 - x_4^8x_6^7 \quad \longleftrightarrow \quad x_1^2x_2^{13}y_4^8x_5^5y_6^7.$$

The following theorem states that the Gröbner basis is a minimal test set for the family of integer programs specified by the matrix A and the cost vector c.

THEOREM 2.3. *Let u be a feasible solution of (1.1). Then u is non-optimal if and only if there exists a vector $g = g^+ - g^-$ in the Gröbner basis $\mathcal{G}_{A,c}$ with $g^+ \leq u$, and in this case the vector $u - g$ is a better feasible solution than u. There is no smaller set than $\mathcal{G}_{A,c}$ which has this property.*

PROOF. By construction, every element $g = g^+ - g^-$ satisfies $c \cdot g = c \cdot g^+ - c \cdot g^- > 0$. The if-direction follows because $(u - g) \cdot c < u \cdot c$ and $g^+ \leq u$ is equivalent to $u - g$ being feasible. For the only-if direction suppose that no $g \in \mathcal{G}_{A,c}$ satisfies $g^+ \leq u$. This means that no element of $\text{Min}(\mathbb{N}^n \backslash \text{Opt}_{A,c})$ lies below u in the poset $\mathbb{N}^n$. But this means that $u \in \text{Opt}_{A,c}$. The minimality of $\mathcal{G}_{A,c}$ holds because every element of $\text{Min}(\mathbb{N}^n \backslash \text{Opt}_{A,c})$ has to be reducible by some vector in the test set. □

Under a certain genericity hypothesis on the matrix A, the elements in the Gröbner basis are in bijection with the *neighbors of the origin*, which is a test set for integer programming introduced by Herbert Scarf [**Sca**]. The connection between neighbors and Gröbner bases was studied in a commutative algebra setting in [**PS**].

Let us assume that the Gröbner basis $\mathcal{G}_{A,c}$ is known to us in some explicit or implicit form. If we are given any feasible solution $u \in \mathbb{N}^n$ then the integer programming problem (1.1) can be solved by the following one-line algorithm:

(2.1) **While** there exists $g \in \mathcal{G}_{A,c}$ with $g^+ \leq u$ **do** replace u by $u - g$.

The problem of constructing a first feasible solution u from the right hand side b can be solved by the same reduction process but for a different Gröbner basis. The idea is completely analogous to *Phase One in the Simplex Algorithm.* To keep our discussion simple, we will assume that some feasible solution u is known beforehand.

One of the objectives of this lecture is to dispel the belief, held by many experts in complexity theory and combinatorial optimization, that the algebraic notion of Gröbner bases is utterly useless when it comes to designing efficient algorithms. Let me begin by pointing out that **computing Gröbner bases is easy and fun.**

My currently favorite tool for producing the Gröbner basis $\mathcal{G}_{A,c}$ from the matrix A and the cost vector c is the software `4ti2` developed by Raymond Hemmecke. It can be found at `www.4ti2.de` and is ridiculously easy to download and run. It took me (= a technologically challenged individual) precisely three minutes to install `4ti2` on my (ancient) computer, and another minute later I was already enjoying my first Gröbner basis on the screen. Actually, I don't recall having ever encountered a piece of mathematical software that was simpler to use than `4ti2`.

The first non-coin example I tried had $d = 3$ and $n = 7$. The matrix A was filled snakewise by prime numbers and the vector c was filled by square integers. The input to `4ti2` consists of a matrix in a file named `example` in the format

```
3 7
 2  3  5  7 11 13 17
43 41 37 31 29 23 19
47 53 59 61 67 71 73
```

and a cost vector in a file `example.cost` in the format

```
1 7
1 4 9 16 25 36 49
```

After typing `groebner example` and hitting the return key, about three seconds later, the Gröbner basis appeared in a new file called `example.gro`:

```
241 7
-6 14 -11 2 1 0 0
-6 13 -9 2 0 -1 1
0 -1 2 0 -1 -1 1
-4 11 -7 -1 -2 3 0
10 -3 -35 38 -1 -7 0
0 22 -53 39 -2 -4 0
4 11 -46 40 0 -7 0
 ... ... ... ...
```

This Gröbner basis consists of 241 vectors in $\mathbb{Z}^7$, and it represents a solution to the parametric problem of minimizing $\sum_{i=1}^{7} i^2 \cdot u_i$ over all vectors $u \in \mathbb{N}^7$ that satisfy

$$(2.2) \qquad \begin{pmatrix} 2 & 3 & 5 & 7 & 11 & 13 & 17 \\ 43 & 41 & 37 & 31 & 29 & 23 & 19 \\ 47 & 53 & 59 & 61 & 67 & 71 & 73 \end{pmatrix} \cdot u \quad = \quad b.$$

Knowing the 241 vectors in the Gröbner basis, we can now apply the reduction algorithm (2.1) starting with any given feasible solution u. Take, for instance, $u = (100, 100, 100, 100, 100, 100)^T$. The corresponding right hand side is

$$b \quad = \quad A \cdot u \quad = \quad \begin{pmatrix} 5800 \\ 22300 \\ 43100 \end{pmatrix}.$$

The algorithm (2.1) reduces u to the optimal solution $u^* = (62, 8, 176, 17, 423, 0, 0)$. The optimal value is found to be $c^* = 12,525$.

Knowledge of the Gröbner basis allows us to answer more advanced structural questions about the system (2.2). One such question is that of finding the *integer programming gap*, a topic to be discussed in Section 4. Another example is the question of *sensitivity analysis* with respect to the cost function. Suppose that the cost vector is allowed to vary in a neighborhood of the given vector c. Then the Gröbner basis $\mathcal{G} = \mathcal{G}_{A,c}$ remains unchanged provided c ranges in the *Gröbner cone*, which is defined by the following linear inequalities in the unknowns $c_1, \ldots, c_n$:

$$(2.3) \qquad c \cdot g > 0 \quad \text{for all } g \in \mathcal{G}.$$

For instance, in our coin example, the Gröbner cone is the set of all solutions to

$$3c_2 + c_4 > 3c_3\,, \quad 6c_2 > 5c_1 + c_4\,, \quad 3c_2 + 4c_3 > 5c_1 + 2c_4\,, \quad 5c_1 + 3c_4 > 8c_3.$$

The collection of all Gröbner cones in $\mathbb{R}^n$ forms the *Gröbner fan* of the matrix A. This is an important invariant which allows us study how the solution of (1.1) changes as both b and c are allowed to vary. See [**ST**] for the basic theory.

The Gröbner fan of a matrix A can be efficiently calculated using the algorithm of Huber and Thomas [**HT**]. A highly efficient implementation was recently given by Anders Jensen in his program `CaTS`. This piece of software can currently be found at the web page `http://www.soopadoopa.dk/anders/cats/cats.html`.

3. Hilbert bases and Graver bases

Gröbner bases are closely related to other natural notions of test sets arising in the theory of integer programming. A classical such notion is that of a *Hilbert basis.* Consider the problem of solving a homogeneous system of linear equations over the non-negative integers. As before, we assume that the defining matrix A has d rows and n columns. Then our solution set is the following semigroup:

$$(3.1) \qquad \ker_{\mathbb{N}}(A) \quad = \quad \big\{\, u \in \mathbb{N}^n \;:\; A \cdot u = 0 \,\big\}.$$

Consider the subset of non-zero minimal elements of the semigroup:

$$(3.2)\;\; \mathcal{H}_A \;=\; \big\{u \in \ker_{\mathbb{N}}(A)\backslash\{0\} : \text{no element } v \in \ker_{\mathbb{N}}(A)\backslash\{0,u\} \text{ satisfies } v \leq u\big\}.$$

The following result is due to the 19th century invariant theorist Paul Gordan:

PROPOSITION 3.1. *The set $\mathcal{H}_A$ is finite. It is the unique minimal set such that every vector in* $\ker_{\mathbb{N}}(A)$ *is an* $\mathbb{N}$*-linear combination of elements in* $\mathcal{H}_A$.

The finite set $\mathcal{H}_A$ is called the *Hilbert basis* of the matrix A. Hilbert bases can also be computed using the program `4ti2`. We consider the same matrix as in (2.2) but we now alternate the sign pattern of its columns in the input file `example`:

```
3 7
  2  -3   5  -7  11  -13  17
 43 -41  37 -31  29  -23  19
 47 -53  59 -61  67  -71  73
```

After typing `hilbert example` and hitting the return key, about six seconds later, the Hilbert basis appears on a new file called `example.hil`:

```
1305 7
4 34 62 38 3 0 1
4 35 64 38 2 1 2
4 60 123 77 1 0 5
4 36 66 38 1 2 3
... ... ... ...
0 673 980 0 2 647 324
0 674 982 0 1 648 325
0 675 984 0 0 649 326
```

The Hilbert basis consists of $1,305$ vectors, and it has a lot of internal structure.

Hilbert bases play an important role in the recent work of Robert Weismantel and his collaborators on "primal methods in integer programming". The paper [**HKW**] introduces the notion of *integral basis* which is a slight generalization of Hilbert bases, and it presents a simplex-like *integral basis algorithm* which is shown to perform very well on standard benchmark problems in integer programming.

A larger test set associated with an integer matrix A is the Graver basis, which can be defined as follows. For any sign pattern $\sigma \in \{-1,+1\}^n$ let D_σ be the $n \times n$-diagonal matrix with i-th entry σ_i. The *Graver basis* of A is the finite set

$$(3.3) \qquad \mathcal{GR}_A \quad := \quad \bigcup_{\sigma \in \{-1,+1\}^n} D_\sigma \cdot \mathcal{H}_{AD_\sigma}$$

In this definition, we are taking the union over the 2^n Hilbert bases for the various matrices $A \cdot D_\sigma$. The signs are adjusted so that each Hilbert basis lies in the kernel of the original matrix A. Proposition 3.1 ensures that the Graver basis $\mathcal{GR}_A$ is a finite subset of $\ker_{\mathbb{Z}}(A)$. The following result is proved in [**Stu**, §7].

PROPOSITION 3.2. *The Graver basis is a universal Gröbner basis. It contains, up to negating vectors, the Gröbner bases of A for all cost functions. In symbols,*

$$\bigcup_{c\in\mathbb{Z}^n} \mathcal{G}_{A,c} \quad\subseteq\quad \mathcal{GR}_A. \tag{3.4}$$

The Graver basis is the ultimate test set one can compute for a given integer matrix A. It provides a parametric solution to the integer programming problem (1.1) when both the right hand side b and the cost function c are allowed to vary.

EXAMPLE 3.3. If A is the 2×4-matrix (1.2) in our coin problem, then

$$\mathcal{GR}_A \;=\; \{\;(0,3,-4,1),\;(-5,6,0,1),\;(-5,3,4,-2),\;(5,0,-8,3),\;(-5,9,-4,0)\}.$$

This Graver basis has only one more element than the Gröbner basis (1.4). The advantage of the Graver basis over the Gröbner basis is that we can now use (2.1) to solve the coin problem with respect to an arbitrary cost vector c. □

The Graver basis has another natural interpretation in integer programming. Consider our original problem (1.1) but now add the requirement that the coordinates u_i of the solution u are bounded above by some quantities a_i.

$$\text{Minimize } c\cdot u \text{ subject to } A\cdot u \,=\, b\,,\;\; u\in\mathbb{N}^n\,,\;\text{ and }\; u\leq a. \tag{3.5}$$

Here we regard A and c as fixed and $(a,b)\in\mathbb{Z}^{n+d}$ as unspecified. It turns out that the Graver basis is the unique minimal test set for this family of integer programs.

THEOREM 3.4. *Let u be a feasible solution of (3.5). Then u is non-optimal if and only if there exists $g\in\mathcal{GR}_A$ with $g^+\leq u$ and $g^-\leq a-u$, and in this case $u-g$ improves u. There is no smaller set than $\mathcal{GR}_A$ which has this property.*

PROOF. We must prove the only if direction. Suppose u is non-optimal for (3.5) and let v be the corresponding optimal solution. Pick $\sigma\in\{-1,+1\}^n$ so that $D_\sigma(v-u)$ is a nonnegative vector. There exist elements $h_1,\ldots,h_r$ (not necessarily distinct) in the Hilbert basis $\mathcal{H}_{D_\sigma A}$ such that $D_\sigma(v-u) = h_1+\cdots+h_r$. Hence

$$v-u \;=\; D_\sigma h_1+\cdots+D_\sigma h_r,$$

where each summand lies in $\mathcal{GR}_A$. Since $c\cdot(v-u)<0$, there exists at least one index i such that $D_\sigma\cdot h_i\cdot c<0$. The vector $g=-D_\sigma\cdot h_i$ lies in $\mathcal{GR}_A$. The construction implies that it satisfies $g^+\leq u$ and $g^- = (D_\sigma\cdot h_i)^+\leq(v-u)^+\leq a-u$.

We now show that every element $g=g^+-g^-$ of $\mathcal{GR}_A$ is needed in a test set for our problem. Suppose that $c\cdot g<0$ and define $a=g^++g^-$ and $b=Ag^+=Ag^-$. With these choices of a and b, the vectors g^+ and g^- are the only two feasible solutions for (3.5). Hence the move from g^+ to g^- must be in the test set. □

In light of Proposition 3.2 and Theorem 3.4, it is highly desirable to be able to precompute the Graver basis of a given integer matrix. An algorithm for this computation is available in `4ti2`. But the reader should be warned that Example 3.3 is somewhat misleading: the Graver basis is often much larger than the Gröbner basis and it takes much longer to compute it. Consider again our `example` matrix,

```
3 7
 2  3  5  7 11 13 17
43 41 37 31 29 23 19
47 53 59 61 67 71 73
```

The command `graver example` produces the Graver basis in a file `example.gra`:

```
29417 7
0 1 2 0 -1 1 1
0 24 57 39 0 -2 2
0 25 59 39 -1 -1 3
 ... ... ... ...
14 -9 10 81 -6 -89 -37
64 86 11 1 -6 -48 -28
114 229 126 -1 -6 -11 -15
124 268 161 -7 -6 10 -6
 ... ... ... ...
```

This Graver basis has 29,417 elements and it took a couple of hours to compute. One nice feature of the Graver basis computation in `4ti2` is that the program allows the exploitation of symmetry. In many applications (e.g. in statistics) there is a group of symmetries acting on the columns of the matrix A, and the Graver basis $\mathcal{GR}_A$ is invariant under these symmetries. This feature allows the computation of some interesting Graver bases whose cardinalities are in the range of one million.

4. The integer programming gap

A commonly used first step towards solving a hard integer programming problem (1.1) is to begin by solving its *linear programming relaxation*:

$$\text{(4.1)} \qquad \text{Minimize } c \cdot u \text{ subject to } A \cdot u = b \text{ and } u \in \mathbb{R}^n_{\geq 0}.$$

Linear programming problems are much easier both in practise and in theory. They can be solved in polynomial time using interior-point methods, and the simplex algorithm performs well in practise. The purpose of this section is to offer algebraic tools for comparing the hard problem (1.1) with the easier problem (4.1). For an algebraic perspective on the linear programming relaxtion see [**HT2**].

As before, we fix $A \in \mathbb{Z}^{d \times n}$ and $c \in \mathbb{Z}^n$ and regard $b \in \mathbb{Z}^d$ as unspecified. We write $\text{IPopt}_{A,c}(b)$ for the optimal value of the integer program (1.1) and we write $\text{LPopt}_{A,c}(b)$ for the optimal value of the corresponding linear program (4.1). The difference of these quantities is a non-negative rational number

$$\text{(4.2)} \qquad \text{IPopt}_{A,c}(b) - \text{LPopt}_{A,c}(b) \quad \geq \quad 0.$$

The *integer programming gap* is defined as the maximum of the differences (4.2) as b ranges over all right hand sides such that (1.1) is feasible:

$$\text{(4.3)} \quad \text{gap}(A,c) \;=\; \max\bigl\{ \text{IPopt}_{A,c}(b) - \text{LPopt}_{A,c}(b) \;:\; b \in \mathbb{Z}^d \text{ feasible for (1.1)} \bigr\}.$$

It appears as if we are taking the maximum over infinitely many different values, one for each feasible b, but actually there are only finitely many possible values for (4.2) if A and c are fixed, so the maximum is attained.

EXAMPLE 4.1. The integer programming gap of the coin problem (1.2) equals

$$\text{gap}(A,c) \;=\; 76/15 \;=\; 5.0666666...$$

This is the maximum advantage to be gained if we allow our coins to be cut into fractional pieces. The gap is attained for the right hand side $b = \binom{10}{114}$ in Example 1.1, where $\text{IPopt}_{A,c}(b) = 6$ is realized by $u = (4,2,0,4)$. The optimal value of the linear program (4.1) is $14/15 = 0.93333...$ and is attained by $u' = (0,0,136/15,14/15)$. Thus the best way to make one dollar and fourteen cents with ten fractional coins is to take $136/15$ dimes and $14/15$ quarters. □

We now give a recipe for computing the gap by solving several auxiliary linear programming problems. For any optimal vector $u \in \mathbb{N}^n$ we define the *increase set*

$$\operatorname{incr}(u) \quad := \quad \{\, i \in \{1,2,\ldots,n\} \,:\, u + e_i \text{ optimal} \,\}.$$

A vector $u \in \mathbb{N}^n$ is said to be *maximally optimal* for (1.1) if $u + a$ is optimal for all vectors $a \in \mathbb{N}^n$ whose support $\{i : a_i > 0\}$ is a subset of the increase set $\operatorname{incr}(u)$. For any fixed maximally optimal $u \in \mathbb{N}^n$, we consider the following linear program:

$$\text{(4.4)} \quad \text{Maximize } c\cdot(u-v) \text{ subject to } A\cdot(u-v) \;=\; 0 \text{ and } v_i \geq 0 \text{ for all } i \notin \operatorname{incr}(u).$$

Here $v = (v_1, \ldots, v_n) \in \mathbb{R}^n$ is an indeterminate real vector. Its coordinates v_i are the decision variables in the linear program (4.4).

THEOREM 4.2. (Hoşten and Sturmfels [**HS2**]) *The maximum of the optimal values of the auxiliary linear programs (4.4), as u ranges over all maximally optimal solutions to (1.1), coincides with the integer programming gap,* $\operatorname{gap}(A, c)$.

EXAMPLE 4.3. Our coin problem (1.2) has three maximally optimal solutions:

$$\text{(4.5)} \qquad (4, 2, \underline{0}, \underline{0})\,,\ (\underline{0}, 2, \underline{0}, 2)\,,\ (\underline{0}, 5, 3, 0).$$

In each case the increase set $\operatorname{incr}(u)$ is indicated by the underlined coordinates. The vectors in (4.5) are easily derived from the Gröbner basis (1.3). For instance, the last portfolio (consisting of five nickels and three dimes) is maximally optimal because adding one nickel, dime or quarter makes that portfolio non-optimal but adding any number of pennies is fine. The program (4.4) for that portfolio equals

$$\text{Maximize } 5 - v_2 - v_4 \text{ subject to } v_1 + v_2 + v_3 + v_4 = 8$$
$$v_1 + 5v_2 + 10v_3 + 25v_4 = 55\,,\ v_2 \geq 0\,, v_3 \geq 0 \text{ and } v_4 \geq 0.$$

The optimal value is 5. The optimal values of (4.4) for $(4, 2, \underline{0}, \underline{0})$ and $(\underline{0}, 2, \underline{0}, 2)$ are 76/15 and 4 respectively, and hence $\operatorname{gap}(A, c) = \max\{4, 5, 76/15\} = 76/15$. □

Theorem 4.2 furnishes an algorithm for computing $\operatorname{gap}(A, c)$ because the set of maximally optimal solutions to (1.1) is always finite and can be computed from the Gröbner basis $\mathcal{G}_{A,c}$. This is done as follows: We first form the monomial ideal M generated by all monomials x^{g^+} where $g = g^+ - g^-$ is in $\mathcal{G}_{A,c}$. Then we apply the algebraic process of *irreducible decomposition* to the monomial ideal M. A vector u is maximally optimal if and only if the irreducible monomial ideal $\langle\, x_i^{u_i+1} : i \notin \operatorname{incr}(u)\,\rangle$ appears among the irreducible components of M.

An efficient implementation of irreducible decomposition of monomial ideals was developed by Alex Milowski in his Master's thesis project at San Francisco State University. His software is posted at `http://www.milowski.com/software.html`. The non-trivial gap computations in Examples 4.4 and 6.3 were done by Hoşten and Milowski using `4ti2` (to derive the Gröbner basis), Milowski's software (to get the maximally optimal solutions) and `maple` (to solve the linear programs (4.4)).

EXAMPLE 4.4. Let $d = 3$, $n = 7$ and consider the instance discussed in (2.2):

$$A = \begin{pmatrix} 2 & 3 & 5 & 7 & 11 & 13 & 17 \\ 43 & 41 & 37 & 31 & 29 & 23 & 19 \\ 47 & 53 & 59 & 61 & 67 & 71 & 73 \end{pmatrix}, \qquad c = (1, 4, 9, 16, 25, 36, 49).$$

There are 553 maximally optimal solutions, and the gap is

$$\text{(4.6)} \qquad \operatorname{gap}(A, c) \quad = \quad 43771/183 \quad = \quad 239.1857923$$

The gap is attained by the right hand side $b = (661, 1710, 3994)^T$. For this choice of b, the optimal value of (1.1) equals 1,757 and is given by the optimal solution $u = (7, 4, 0, 22, 0, 3, 26)$, while the optimal value of (4.1) is a little less than 1,518 and is given by the optimal solution $u = (0, 0, 0, 0, 14029/244, 463/366, 521/732)$. □

5. Short rational generating functions

The importance of rational generating functions for lattice point problems has been known to combinatorialists for a long time. Their role as an efficient tool in integer programming, however, has been recognized only quite recently, in response to the polynomial time algorithms of Barvinok [**Bar**] and Barvinok-Woods [**BW**]. This work was further extended by De Loera et.al. [**DHHHSY**], [**DHHHY**]. This section reports on these methods and their implementations in the software `LattE`.

As a point of entry consider the following variant of our problem: *List all optimal solutions to the integer program* (1.1). For a concrete example take $d = 1$, $n = 4$, $A = (\,1\;\,1\;\,1\;\,1\,)$, $c = (0, 0, 0, 1)$, and suppose $b \gg 0$. Here (1.1) equals

$$\text{(5.1)} \qquad \text{Minimize } u_4 \text{ subject to } u_1 + u_2 + u_3 + u_4 \;=\; b \text{ and } u_1, u_2, u_3, u_4 \in \mathbb{N}.$$

The set of optimal solutions is the set of all lattice points $(u_1, u_2, u_3, 0)$ in a large triangle. We can write them all down as the terms of the generating function

$$\text{(5.2)} \qquad \sum_{u \text{ optimal for } (5.1)} x_1^{u_1} x_2^{u_2} x_3^{u_3} x_4^{u_4} \quad = \quad \sum_{u_1=0}^{b} \sum_{u_2=0}^{b-u_1} x_1^{u_1} x_2^{u_2} x_3^{b-u_1-u_2}.$$

The number of terms in this series equals $(b+1)(b+2)/2 = O(b^2)$. This quantity is exponential in the size of the input, which is $O(\log(b))$. Indeed, the number of bits needed to write down the line (5.1) grows like the logarithm of the integer b, while the number of terms on the right hand side of (5.2) is exponential in $\log(b)$. It appears to be impossible to "list" all feasible solutions to (5.1) in polynomial time, given that their number grows exponentially in the input size. Nonetheless, it can be done, namely, by rewriting (5.2) as the *short rational generating function*

$$x_1^b \cdot (1 - \frac{x_2}{x_1})^{-1} (1 - \frac{x_3}{x_1})^{-1} \;+\; x_2^b \cdot (1 - \frac{x_1}{x_2})^{-1} (1 - \frac{x_3}{x_2})^{-1} \;+\; x_3^b \cdot (1 - \frac{x_1}{x_3})^{-1} (1 - \frac{x_2}{x_3})^{-1}.$$

The reader is invited to check that this rational function equals the series (5.2). The rational function can be computed in time $O(\log(b))$ and it represents the "list" of all optimal solutions to (5.1). This approach works for any integer program:

THEOREM 5.1. *Suppose that d and n are fixed. Then the number of optimal solutions to* (1.1) *and the rational generating function* $\sum \{\, x^u \;:\; u \text{ optimal for } (1.1) \,\}$, *which encodes the set of optimal solutions, can be computed in polynomial time.*

PROOF. The optimal value c^* of (1.1) can be computed in polynomial time using Lenstra's algorithm [**Len**]. Now apply Barvinok's lattice point algorithm [**Bar**] to the polytope $\{\, u \in \mathbb{R}^n_{\geq 0} \;:\; Au = b,\; c \cdot u = c^* \,\}$. It computes the desired generating function and its evaluation at $(1, 1, \ldots, 1)$ in polynomial time. □

The techniques underlying Barvinok's algorithm were developed substantially further by Barvinok and Woods [**BW**]. Using their Projection Theorem, one can derive polynomial-time algorithms based on rational generating functions for essentially all of the algorithmic questions we have encountered so far. We refer to [**BW**] and [**DHHHSY**] for proofs of various parts of the following theorem.

THEOREM 5.2. *Consider a matrix* $A \in \mathbb{Z}^{d\times n}$ *and a vector* $c \in \mathbb{Z}^d$ *whose dimensions* d *and* n *are fixed. Then the rational generating functions which encode the following sets can be computed in time polynomial in the bit complexity of* A *and* c*:*

(1) *the Gröbner basis* $\mathcal{G}_{A,c}$,
(2) *the set* $\mathrm{Opt}_{A,c}$ *of all optimal solutions,*
(3) *the set* $\mathrm{Min}(\mathbb{N}^n \backslash \mathrm{Opt}_{A,c})$ *of minimally non-optimal points,*
(4) *the Hilbert basis* $\mathcal{H}_A$,
(5) *the Graver basis* $\mathcal{GR}_A$,
(6) *the set of maximally optimal solutions, and*
(7) *the integer programming gap* $\mathrm{gap}(A, c)$.

The result (7) about the gap appears in [**HS2**]. The objects in (1)-(6) are highly structured subsets of $\mathbb{Z}^n$. It is this special structure which allows for a short encoding. For encoding the Gröbner basis $\mathcal{G}_{A,c}$, the paper [**DHHHSY**] uses a generating function in $2n$ variables as in (1.5). But all the sets in (1)-(6) can also be coded as formal sums of Laurent monomials (representing vectors in $\mathbb{Z}^n$), and the Barvinok-Woods method will give short rational functions for these encodings.

A magnificent computer program for solving lattice point problems by means of short rational generating functions has been developed by the group of Jesus De Loera at UC Davis. It is called `LattE` and can be obtained at the web site `http://www.math.ucdavis.edu/~latte/`. This program can be used to count the number of feasible solutions to an integer program (1.1) as follows.

Consider our coin problem in (1.2) with $b = \binom{999}{5000}$, so we wish to arrange 999 coins to be worth fifty dollars. In order to determine in how many ways this can be accomplished, we create the following `LattE` input file which we call `coins`:

```
6 5
999 -1  -1  -1  -1
5000 -1 -5 -10 -25
 0    1  0   0   0
 0    0  1   0   0
 0    0  0   1   0
 0    0  0   0   1
2 1 2
```

The command `latte equ coins` will count the number of feasible solutions to (1.1). The output which appears on the screen reveals that the answer is $9,352$:

```
This is LattE v1.0 beta.   (September 17, 2002)
Revised version.           (Aug        1, 2003)
The polytope has 4 vertices.
 ....  ....  .... ...
Creating generating function.
Starting final computation.

  **** THE GRAND TOTAL IS: 9352 ****

Computation done.
Time: 0.01 sec
```

This run of `LattE` has created the following output on a new file called `coins.maple`:

```
gF:=x[0]^4999170*x[1]^(-8000506)*x[2]^1000*x[3]^166/((1-x[0]^(-4995)
```

```
*x[1]^7993*x[2]^(-1)*x[3]^(-1))*(1-x[0]^5*x[1]^(-9)*x[3]^(-1)))+
x[0]^4999170*x[1]^(-8000506)*x[2]^1000*x[3]^166/((1-x[0]^(-9985)
 ....   ....    ....   ....    ....    ....    ....    ....
... *x[1]^7993*x[2]^(-1)*x[3]^(-1))*(1-x[0]^(-5)*x[1]^9*x[3]));
```

This is the short rational generating function representing the formal sum of $9,352$ monomials, one for each feasible solution. You can get the expanded form of this generating function reading the file `coins.maple` into the computer algebra system `maple`. After you have done this, please type the `maple` command `simplify(gF);`.

The program `LattE` can also be used to solve the minimization problem (1.1). To this end we need to add the cost vector `0 1 0 1` in an extra line at the end of the input file `coins`. Typing now the `LattE` command sequence `./latte equ min coins2`, we obtain the following output on the screen

```
 ....  ...    ....    ....
All cones have been decomposed.
6 cones in total.
Computing the points in the Parallelepiped of the unimodular Cones.
An optimal solution for [0 1 0 1] is: [555 2 441 1].
The optimal value is: 3.
The gap is: 3
Computation done.
```

We conclude from this `LattE` session that the best way of making fifty dollars with 999 coins is to take one quarter, two nickels, 441 dimes and 555 pennies.

I tried `LattE` on considerably bigger problems and I found that it performs quite well. The speed is particularly impressive for knapsack problems (d = 1) with large integer coefficients. For this class of problems, `LattE` is faster than the current version of the commerical software `CPLEX` on some instances. This parallels the observation, already made in [**HS1**], that programs like `CPLEX` are not always the best choice for low-dimensional problems with large integers, given that they are designed for highly structured 0/1 problems with many variables.

The authors of `LattE` informed me that they intend to incorporate all of the tasks listed in Theorem 5.2 into a future version of their program. The lesson to be learned here is that algebraic software like `4ti2` and `LattE` can definitely play a useful role in the box of tools available to practitioners of integer programming.

6. Some integer programs arising in statistics

We present an application to the statistical theory of *disclosure limitation*. See [**CG**] and [**DF**] and the references therein. Suppose we are given data in the form of an n-dimensional table of nonnegative integers. The aim is to release some marginals of the table but not the table's entries themselves. If the range of possible values that a particular entry can attain in any table satisfying the released marginals is too narrow then this entry may be exposed. This shows the importance of determining tight upper and lower bounds for each entry in a given table.

A choice of marginals corresponds to fixing subsets $F_1, \ldots, F_k$ of $\{1, \ldots, n\}$. It can be represented by a zero-one matrix A, as described in [**HSu**, §1]. In statistical language, the matrix A specifies a *hierarchical model* for a *contingency table* with n *factors*. Suppose v is a table with nonnegative integer entries, where the marginals are computed according to a fixed hierarchical model A and let $v_{i_1 i_2 \cdots i_n}$ be a particular cell of the table v. What we are interested in is the following **table entry**

security problem: *Compute optimal lower and upper bounds L and U such that $L \leq u_{i_1 i_2 \cdots i_n} \leq U$ for all tables u which have the same marginals as v.*

The table entry security problem is an integer program: minimize (or maximize) $u_{i_1 i_2 \cdots i_n}$ over all tables with nonnegative *integer* entries subject to fixing the marginals. In order to write this integer program in the standard form (1.1), we need to give the precise definition of the relevant matrices A. Consider $d_1 \times \cdots \times d_n$-tables with entries $u_{i_1 i_2 \cdots i_n}$ where $1 \leq i_j \leq d_j$. We fix a hierarchical model by specifying $F_1, \ldots, F_k$. The marginals of our table are computed with respect to these subsets. If $F_i = \{j_1, \ldots, j_s\}$ then the F_i-marginal is a $d_{j_1} \times \cdots \times d_{j_s}$ table b with entries

$$b_{k_1 \cdots k_s} \quad = \sum_{i_{j_1}=k_1, \ldots, i_{j_s}=k_s} u_{i_1 \cdots i_n}. \tag{6.1}$$

We define A to be the zero-one matrix with $d_1 d_2 \cdots d_n$ columns representing the linear map that computes the marginals of tables. We let u be the vector of variables representing the cell entries. Then $A \cdot u$ represents the k lower-dimensional tables computed as in (6.1). The **table entry security problem** is

$$\text{Minimize (Maximize) } u_{11\cdots 1} \quad \text{subject to} \quad A \cdot u = b,\ u \geq 0,\ u \text{ integral.} \tag{6.2}$$

Here we only consider the cell entry $u_{11\cdots 1}$ (corresponding to the first column of A) because there is a transitive symmetry group acting on the columns of A.

EXAMPLE 6.1. The classical *transportation problem* [**Sch**, p. 221] corresponds to $d_1 \times d_2$-tables where the marginals are computed with respect to $F_1 = \{1\}$ and $F_2 = \{2\}$. The *three-dimensional transportation problem* [**Vla**] concerns $d_1 \times d_2 \times d_3$-tables with $F_1 = \{1, 2\}$, $F_2 = \{1, 3\}$, and $F_3 = \{2, 3\}$. The marginals are

$$b_{ij} = \sum_k u_{ijk}\,, \quad b_{ik} = \sum_j u_{ijk}\,, \quad b_{jk} = \sum_i u_{ijk}.$$

For a discussion from the Gröbner basis perspective see [**Stu**, §14.C]. □

EXAMPLE 6.2. Consider the *four-cycle model* for binary random variables. Here $n = 4$, $d_1 = d_2 = d_3 = d_4 = 2$, $F_1 = \{1, 2\}$, $F_2 = \{2, 3\}$, $F_3 = \{3, 4\}$, and $F_4 = \{1, 4\}$. The matrix A has $d_1 d_2 + d_2 d_3 + d_3 d_4 + d_1 d_4 = 16$ rows and it has $d_1 d_2 d_3 d_4 = 16$ columns. We write it in `4ti2` format on a file name `fourcycle`:

```
16
16
 1 1 1 1 0 0 0 0 0 0 0 0 0 0 0 0
 0 0 0 0 1 1 1 1 0 0 0 0 0 0 0 0
 0 0 0 0 0 0 0 0 1 1 1 1 0 0 0 0
 0 0 0 0 0 0 0 0 0 0 0 0 1 1 1 1
 1 1 0 0 0 0 0 0 1 1 0 0 0 0 0 0
 0 0 1 1 0 0 0 0 0 0 1 1 0 0 0 0
 0 0 0 0 1 1 0 0 0 0 0 0 1 1 0 0
 0 0 0 0 0 0 1 1 0 0 0 0 0 0 1 1
 1 0 0 0 1 0 0 0 1 0 0 0 1 0 0 0
 0 1 0 0 0 1 0 0 0 1 0 0 0 1 0 0
 0 0 1 0 0 0 1 0 0 0 1 0 0 0 1 0
 0 0 0 1 0 0 0 1 0 0 0 1 0 0 0 1
 1 0 1 0 1 0 1 0 0 0 0 0 0 0 0 0
 0 1 0 1 0 1 0 1 0 0 0 0 0 0 0 0
```

```
0 0 0 0 0 0 0 0 1 0 1 0 1 0 1 0
0 0 0 0 0 0 0 0 0 1 0 1 0 1 0 1
```

The Graver basis of this matrix A consists of 106 vectors. The `4ti2` command `graver fourcycle` delivers the Graver basis on a new file `fourcycle.gra`:

```
106 16
0 0 0 0 0 1 0 -1 1 -1 0 0 -1 0 0 1
-1 0 1 0 1 0 -1 0 1 0 0 -1 -1 0 0 1
0 0 0 0 0 1 0 -1 0 0 0 0 0 -1 0 1
0 0 0 0 1 0 -1 0 0 0 0 0 -1 0 1 0
1 0 -1 0 0 0 0 0 -1 0 1 0 0 0 0 0
0 0 -1 1 0 0 1 -1 0 0 0 0 0 0 0 0
-1 1 0 0 1 -1 0 0 0 0 0 0 0 0 0 0
0 -1 1 0 0 1 -1 0 1 0 0 -1 -1 0 0 1
 ... ... ... ...  ... ... ... ...
```

In light of Proposition 3.2, we can use this Graver basis with the Algorithm (2.1) to solve (1.1) for any cost function. In particular, we can use it solve (6.2). □

As the parameters $d_1, \ldots, d_n$ increase, it becomes harder to solve the integer program (6.2) exactly. Researchers in disclosure limitation have resorted to solving the linear programming relaxation (4.1) instead: minimize (or maximize) $u_{i_1 i_2 \cdots i_n}$ over all tables with nonnegative *real* entries subject to fixing the marginals. This relaxation is tractable, but it usually fails to deliver the exact integers L and U. One faces the problem of finding the integer programming gap for the table entry security problem. This application was the original motivation for the paper [**HS2**].

EXAMPLE 6.3. What follows may serve as a test case for future software for computing the gap. We consider the K_5-model for five binary random variables. Here $n = 5$, $k = 10$, $d_1 = \cdots = d_5 = 2$ and the F_i are the ten two-element subsets of $\{1,2,3,4,5\}$. The cost function is $c = -e_{11111}$, corresponding to maximizing in (6.2). The matrix A has 40 rows and 32 columns, and it has rank 16. We found

$$\text{gap}(A, c) \quad = \quad 3. \tag{6.3}$$

The gap is attained by the following $2 \times 2 \times 2 \times 2 \times 2$-table:

$$u \quad = \quad e_{11112} + e_{11121} + 2 \cdot e_{11211} + 2 \cdot e_{12111} + 2 \cdot e_{21111} + e_{22222}. \tag{6.4}$$

This table is optimal for the maximization problem in (6.2). The 40 entries of the right hand side vector b are the entries in the ten marginal 2×2-tables:

$$\begin{pmatrix} u_{11***} & u_{12***} \\ u_{21***} & u_{22***} \end{pmatrix} = \begin{pmatrix} 4 & 2 \\ 2 & 1 \end{pmatrix}, \ \ldots\ldots, \ \begin{pmatrix} u_{***11} & u_{***12} \\ u_{***21} & u_{***22} \end{pmatrix} = \begin{pmatrix} 5 & 1 \\ 2 & 1 \end{pmatrix}$$

Since the unit table e_{11111} does not appear in (6.4), we have $\text{IPopt}_{A,c}(b) = 0$.

The optimal value of the linear programming relaxation equals $\text{LPopt}_{A,c}(b) = 3$. This value is attained by the following fractional $2 \times 2 \times 2 \times 2 \times 2$-table

$$\begin{aligned} v \quad = \quad & 3 \cdot e_{00000} + \tfrac{2}{3} e_{11211} + \tfrac{1}{3} e_{11221} + \tfrac{2}{3} e_{12111} + \tfrac{1}{3} e_{12122} + \tfrac{1}{3} e_{12211} + \tfrac{1}{3} e_{12221} + \\ & \tfrac{1}{3} e_{12222} + \tfrac{2}{3} e_{21112} + \tfrac{1}{3} e_{21121} + \tfrac{2}{3} e_{21211} + \tfrac{1}{3} e_{21222} + \tfrac{2}{3} e_{22111} + \tfrac{1}{3} e_{22121}. \end{aligned}$$

We invite the reader to check that the tables u and v have the same marginals. □

References

[Bar] A. Barvinok, *A polynomial time algorithm for counting integral points in polyhedra when the dimension is fixed*, Math. Oper. Res. **19** (1994), no. 4, 769–779.

[BW] A. Barvinok and K. Woods, *Short rational generating functions for lattice point problems*, J. Amer. Math. Soc. **16** (2003), 957-979.

[CT] P. Conti and C. Traverso: Buchberger algorithm and integer programming, *Applied Algebra, Algebraic Algorithms and Error-Correcting Codes* (New Orleans, LA, 1991), 130–139, Lecture Notes in Comput. Sci., 539, Springer, Berlin, 1991.

[CG] L. Cox and J. George, *Controlled rounding for tables with subtotals*, Annals of Operations Research **20** (1989) 141–157.

[DF] A. Dobra and S. Fienberg: *Bounds for cell entries in contingency tables given marginal totals and decomposable graphs*, Proc. Natl. Acad. Sci. USA **97** (2000) 11885–11892.

[DHHHY] J. De Loera, D. Haws, R. Hemmecke, P. Huggins and R. Yoshida, *Three kinds of integer programming algorithms based on Barvinok's rational functions*, IPCO X, Integer Programming and Combinatorial Optimization, Columbia University, New York, July 2004.

[DHHHSY] J. De Loera, D. Haws, R. Hemmecke, P. Huggins, B. Sturmfels and R. Yoshida, *Short rational functions for toric algebra and applications*, J. Symbolic Computation, to appear.

[HKW] U-U. Haus, M. Köppe and R. Weismantel, *The integral basis method for integer programming*. Math. Methods Oper. Res. **53** (2001) 353–361.

[HS1] S. Hoşten and B. Sturmfels: *GRIN: an implementation of Gröbner bases for integer programming*. Integer programming and combinatorial optimization (Copenhagen, 1995), 267–276, Lecture Notes in Comput. Sci., 920, Springer, Berlin, 1995.

[HS2] S. Hoşten and B. Sturmfels: *Computing the integer programming gap*, Combinatorica, to appear.

[HT1] S. Hoşten and R. Thomas: *Gröbner bases and integer programming*, Gröbner bases and applications (Linz, 1998), 144–158, London Math. Soc. Lecture Note Ser., 251, Cambridge Univ. Press, Cambridge, 1998.

[HT2] S. Hoşten and R. Thomas: *Standard pairs and group relaxations in integer programming* Effective methods in algebraic geometry (Saint-Malo, 1998). J. Pure Appl. Algebra **139** (1999), no. 1-3, 133–157.

[HSu] S. Hoşten and S. Sullivant: *Gröbner bases and polyhedral geometry of reducible and cyclic models*, J. Combinatorial Theory, Ser. A **100** (2002) 277–301.

[HT] B. Huber and R. Thomas: *Computing Gröbner fans of toric ideals*, Experimental Mathematics **9** (2000) 321-331.

[Len] H.W. Lenstra, *Integer programming with a fixed number of variables*, Math. Oper. Res. **8** (1983), no. 4, 538–548.

[PS] I. Peeva and B. Sturmfels, *Generic lattice ideals*, J. Amer. Math. Soc. **11** (1998) 263–273.

[Sca] H. Scarf, *Test sets for integer programs*, Lectures on mathematical programming (ismp97) (Lausanne, 1997). Math. Programming 79 (1997), no. 1-3, Ser. B, 355–368.

[Sch] A. Schrijver, *Theory of Linear and Integer Programming*, Wiley Interscience, 1986.

[Stu] B. Sturmfels, *Gröbner Bases and Convex Polytopes*, American Mathematical Society, University Lecture Series, Volume 8, 1996.

[ST] B. Sturmfels and R. Thomas, *Variation of cost functions in integer programming*, Mathematical Programming **77** (1997) 357–387.

[Tho] R. Thomas, A geometric Buchberger algorithm for integer programming. Math. Oper. Res. **20** (1995) 864–884.

[Tho2] R. Thomas, *Gröbner bases in integer programming*, Handbook of combinatorial optimization, Vol. 1, 533–572, Kluwer Acad. Publ., Boston, MA, 1998.

[Vla] M. Vlach, *Conditions for the existence of solutions of the three-dimensional planar transportation problem*, Discrete Appl. Math. **13** (1986), no. 1, 61–78.

Department of Mathematics, University of California, Berkeley, California, 94720
E-mail address: `bernd@math.berkeley.edu`

Proceedings of Symposia in Applied Mathematics
Volume **61**, 2004

Nonlinear and Semidefinite Programming

Stephen J. Wright

ABSTRACT. Nonlinear and semidefinite programming are two areas of optimization that have seen renewed vitality in recent years. Analytical and geometrical techniques provide the mathematical underpinnings for the optimality theory and for the design of algorithms in both areas. This note outlines these topics, with an emphasis on recent developments.

1. Introduction

Nonlinear programming and semidefinite programming are two areas of optimization with a wide variety of practical applications. Both have roots in the early development of optimization and both remain vital research areas today.

The development of theory and algorithms for nonlinear programming has long figured prominently on the agenda of optimization research. Interest in various types of algorithms has waxed and waned with the years, as experience is gained with ever more challenging practical applications and as new software implementations are tried. The renewed interest in interior-point algorithms during the past ten years has been partly responsible for revitalizing interest in the field. New classes of applications such as those with equilibrium constraints has also played a role.

Interior-point approaches have played an even more dramatic role in the explosion of interest in semidefinite programming since 1990. After the success of these approaches for linear and convex quadratic programming, the focus turned squarely to semidefinite programming in about 1994 as researchers realized that interior-point methods provided a theoretically attractive and practical means of solving these problems.

This note reviews nonlinear and semidefinite programming, with an emphasis on the mathematical underpinnings of these fields and on the major algorithmic approaches of current interest. Many new preprints in these areas, and in other areas of optimization, can be found on the following web site:

`http://www.optimization-online.org`

1991 *Mathematics Subject Classification.* Primary 90C30, 90C22, 65K05.
Key words and phrases. Nonlinear Programming, Semidefinite Programming.
Research of the author was supported in part by NSF Grants #MTS-0086559, #ACI-0196485, and #EIA-0127857.

2. Nonlinear Programming: Optimality Theory

Linear programming—minimization of a linear function of real variables subject to constraints on those variables—became a serious topic for mathematical investigation during the 1940s, partly in response to wartime logistical and planning problems. Despite initial skepticism, linear models proved to have widespread application to many practical problems, even those in which the underlying processes were in reality nonlinear. Further impetus was given to the area by Dantzig's discovery of the simplex algorithm [**Dan63**], which proved to be surprisingly effective in practice.

Still, there was persistent interest in nonlinear programming problems (those for which the objective and constraints may be nonlinear), as the linear approximation proved to be inadequate or nonphysical for many applications. Starting in the mid-1950's, a number of different algorithmic approaches were proposed. Theoretical and practical properties of the original algorithms continue to be investigated intensively, and the emergence of new interior-point algorithms and software in the late 1990s has given added momentum to the area. The area continues to be influenced strongly by developments in other areas of optimization, numerical analysis, and applications. For example, developments in iterative methods and preconditioners for large sparse linear algebra problems have given rise to algorithms and codes that can best take advantage of these solvers, and the recent interest in optimization problems in which the constraints arise from differential equations has led to investigation of methods that are best suited to these types of problems.

2.1. Formulation and Solutions. The nonlinear programming problem can be formulated algebraically as follows:

$$\min_x f(x) \qquad \text{subject to } c(x) \geq 0, \;\; d(x) = 0, \tag{2.1}$$

where $x \in \mathbf{R}^n$ is a vector of real variables, $f : \mathbf{R}^n \to \mathbf{R}$ is an objective function, $c : \mathbf{R}^n \to \mathbf{R}^m$ is a vector function of inequality constraints, and $d : \mathbf{R}^n \to \mathbf{R}^p$ is a vector function of equality constraints. The functions f, c, and d are assumed to be smooth, usually twice Frechét differentiable. (We assume the required degree of smoothness in all subsequent results.) The individual components of x, c, and d are denoted by subscripts; for example, c_i represents the ith component of the vector function c. An inequality constraint c_i is said to be *active* at a point x if $c_i(x) = 0$ and *inactive* if $c_i(x) > 0$. The set of all $i \in \{1, 2, \ldots, m\}$ such that $c_i(x) = 0$ is denoted by $\mathcal{A}(x)$.

We say that a point x is *feasible* if it satisfies the constraints; that is, $c(x) \geq 0$ and $d(x) = 0$. The set of all feasible points is denoted by Ω; that is,

$$\Omega = \{x \,|\, c(x) \geq 0, \;\; d(x) = 0\}.$$

Using this notation we can write the nonlinear program in a slightly more abstract form as follows:

$$\min_x f(x) \qquad \text{subject to } x \in \Omega. \tag{2.2}$$

The point x^* is a *global solution* to the problem (2.1) (or (2.2)) if x^* is feasible and if $f(x) \geq f(x^*)$ for all $x \in \Omega$. We say that x^* is a *local solution* if there is an open neighborhood $\mathcal{N}$ of x^* such that $f(x) \geq f(x^*)$ for all $x \in \Omega \cap \mathcal{N}$. The point is a *strict local solution* if $f(x) > f(x^*)$ for all $x \in \Omega \cap \mathcal{N}$ with $x \neq x^*$. The point is an *isolated local solution* if it is the only local solution in $\Omega \cap \mathcal{N}$.

2.2. Optimality Conditions. The first step in devising algorithms for finding solutions of (2.1) is to find a practical means of recognizing solutions. Optimality conditions for nonlinear programming involve a fascinating interplay of the algebraic and geometric representations of the feasible set in the neighborhood of the point in question. As in many other aspects of nonlinear programming, an important role is played by Taylor's theorem, which constructs approximations to a function based on its lower order derivatives.

Constraint qualifications ensure that a linearization of the constraints yields an adequate representation of the feasible set in the vicinity of the point in question. At a point x^*, linearization of the active constraints yields the following set:

$$\mathcal{F} \stackrel{\text{def}}{=} \{s \mid \nabla c_i(x^*)^T s \geq 0, \text{ for all } i \in \mathcal{A}(x^*), \quad \nabla d(x^*)^T s = 0\}.$$

Constraint qualifications ensure that there is some open neighborhood $\mathcal{N}$ of x^* such that the sets $(x^* + \mathcal{F}) \cap \mathcal{N}$ and $\Omega \cap \mathcal{N}$ are qualitatively similar and close together in a certain sense. We illustrate the concept with a few examples.

EXAMPLE 2.1. The following pair of inequality constraints in $\mathbb{R}^2$ yields $\Omega = \{0\}$:

$$c_1(x) = 1 - x_1^2 - (x_2 - 1)^2 \geq 0, \quad c_2(x) = 1 - x_1^2 - (x_2 + 1)^2 \geq 0.$$

At $x^* = (0,0)$, the set $\mathcal{F}$ is defined by $\mathcal{F} = \{(s_1, s_2) \mid s_2 = 0\}$. Hence, for any choice of neighborhood $\mathcal{N}$, the set $\Omega \cap \mathcal{N}$ is a point, whereas $(x^* + \mathcal{F}) \cap \mathcal{N}$ is an open line segment. These two sets are qualitatively different, and constraint qualifications are not satisfied.

EXAMPLE 2.2. The following pair of inequality constraints in $\mathbb{R}^2$ yields a feasible set with nonempty interior in $\mathbb{R}^2$:

$$c_1(x) = 1 - x_1^2 - (x_2 - 1)^2 \geq 0, \quad c_2(x) = 1 - (x_1 - 1)^2 - x_2^2 \geq 0.$$

At $x^* = (0,0)$, the set $\mathcal{F}$ is defined by $\mathcal{F} = \{(s_1, s_2) \mid s_1 \geq 0, \ s_2 \geq 0\}$. For $\mathcal{N}$ sufficiently small, the sets $\Omega \cap \mathcal{N}$ and $(x^* + \mathcal{F}) \cap \mathcal{N}$ differ only in that two of the boundaries of the latter set are straight lines instead of being slightly curved. Hence, the linearized approximation $(x^* + \mathcal{F})$ is locally an adequate representation of Ω, and constraint qualifications are satisfied.

Perhaps the simplest constraint qualification is the assumption that the following set of vectors is linearly independent:

$$\{\nabla c_i(x^*) \mid i \in \mathcal{A}(x^*)\} \ \cup \ \{\nabla d_i(x^*) \mid i = 1, 2, \ldots, p\}. \tag{2.3}$$

This is known as the *linear independence constraint qualification* (LICQ).

First-order optimality conditions are those that depend on the first derivatives of the objective and constraint functions. The following result was discovered independently by Karush [**Kar39**] and Kuhn and Tucker [**KT51**]. (For a proof, see for example Nocedal and Wright [**NW99**].)

THEOREM 2.3. *Suppose that x^* is a local solution of (2.1) and that the LICQ is satisfied at x^*. Then there are vectors $\lambda^* \in \mathbb{R}^m$ and $\gamma^* \in \mathbb{R}^p$ such that the* *following conditions are satisfied:*

$$\nabla f(x^*) - \nabla c(x^*)\lambda^* - \nabla d(x^*)\gamma^* = 0, \tag{2.4a}$$

$$d(x^*) = 0, \tag{2.4b}$$

$$c(x^*) \geq 0, \tag{2.4c}$$

$$\lambda^* \geq 0, \tag{2.4d}$$

$$(\lambda^*)^T c(x^*) = 0. \tag{2.4e}$$

The conditions (2.4) are known as the *Karush-Kuhn-Tucker* or *KKT* conditions; they are the *first-order necessary conditions* for x^* to be a solution when a constraint qualification holds. The condition (2.4e) is known as a complementarity condition; it indicates that for each $i = 1, 2, \ldots, m$, either λ_i^* or $c_i(x^*)$ (or both) are zero.

A variant on the KKT conditions was proposed (independently) by John [**Joh48**]. These so-called *Fritz John* conditions differ from (2.4) only in that a scalar multiplier $\gamma_0^* \geq 0$ is introduced for the term $\nabla f(x^*)$ in (2.4a), and a restriction that not all components of $(\gamma_0^*, \lambda^*, \gamma^*)$ are zero is enforced. It can be shown that any local solution of (2.1) satisfies the Fritz John conditions; no constraint qualification is needed. Note however that when the Fritz John conditions are satisfied with $\gamma_0^* = 0$, the constraint gradients cannot satisfy LICQ. When the Fritz John conditions are satisfied with $\gamma_0^* > 0$, then the KKT conditions are also satisfied. (We simply replace λ^* and γ^* by λ^*/γ_0^* and γ^*/γ_0^*, respectively.)

The *Lagrangian* function is a linear combination of the objectives and constraints, in which the coefficients are the components of the vectors $\lambda \in \mathbf{R}^m$ and $\gamma \in \mathbf{R}^p$ introduced in Theorem 2.4, and known as *Lagrange multipliers*. We denote this function by $\mathcal{L}$ and define it as follows:

$$\mathcal{L}(x, \lambda, \gamma) = f(x) - \lambda^T c(x) - \gamma^T d(x). \tag{2.5}$$

Using this definition, the first KKT condition (2.4a) can be written as

$$\nabla_x \mathcal{L}(x^*, \lambda^*, \gamma^*) = 0. \tag{2.6}$$

A point satisfying the KKT conditions is sometimes known as a *KKT point* or *stationary point*. We can ensure that such a point is in fact a local solution by making additional assumptions on the second derivatives of the Lagrangian. We first define the *critical cone* associated with a KKT point x^* and its corresponding Lagrange multipliers as follows:

$$\mathcal{C} = \left\{ s \;\middle|\; \begin{array}{ll} \nabla c_i(x^*)^T s = 0, & \text{for } i \in \mathcal{A}(x^*) \text{ with } \lambda_i^* > 0 \\ \nabla c_i(x^*)^T s \geq 0, & \text{for } i \in \mathcal{A}(x^*) \text{ with } \lambda_i^* = 0 \\ \nabla d(x^*)^T s = 0 & \end{array} \right\}. \tag{2.7}$$

The critical cone contains those directions from $\mathcal{F}$ for which the first derivatives alone are not sufficient to determine whether the objective function increases or not. In the following theorem the second derivatives are used as a "tiebreaker;" when the second derivatives indicate an increase along these directions as well, we can assert that x^* is a local minimizer.

THEOREM 2.4. *Suppose that the KKT conditions (2.4) are satisfied at x^* for some Lagrange multiplier vectors λ^* and γ^*, and that for all $s \in \mathcal{C}$ with $s \neq 0$ we have*

$$s^T \nabla_{xx}^2 \mathcal{L}(x^*, \lambda^*, \gamma^*) s > 0. \tag{2.8}$$

Then x^* is a strict local solution of (2.1).

(Again, a proof can be found in [**NW99**].) Note that a constraint qualification is not needed for this result. The KKT conditions together with (2.8) are known as *second-order sufficient conditions.*

Finding global solutions of (2.1) is a computationally intractable problem in general. Most algorithms for nonlinear programming aim only to identify local solutions. In a number of practical problems, it is necessary (or highly desirable) to find the global minimizer; molecular conformation and protein folding problem are two such examples. In these cases, special-purpose global minimization algorithms must be designed that take advantage of the nature of the problem in question—but even these are not guaranteed to find the global solution. We do not address the global optimization problem further in these notes.

2.3. Min-max Formulation using the Lagrangian. We can reformulate the problem (2.1) as the following min-max problem:

$$\min_{x} \max_{\lambda \geq 0, \gamma} \mathcal{L}(x, \lambda, \gamma). \tag{2.9}$$

It is easy to see that the max-function $\max_{\lambda \geq 0, \gamma} \mathcal{L}(x, \lambda, \gamma)$ has value ∞ if any of the constraints are violated, and value $f(x)$ if the constraints are satisfied. Hence, the max-function is too nonsmooth to allow a minimization algorithm to be applied directly. A modification to this formulation does however lead to a practical algorithm, as we see in Section 3.3

2.4. Convexity. An important special case of nonlinear programming is *convex programming*, which involves the minimization of a convex function over a convex feasible region. The problem (2.1) has this form if f is convex, c_i are concave for $i = 1, 2, \ldots, m$, and d_i are linear for $i = 1, 2, \ldots, p$. In this case, the global optimization problem is comparatively easy: All local solutions are global solutions, as the following elementary result shows.

THEOREM 2.5. *Suppose that (2.1) is a convex programming problem. Then if x^* is a local solution, it is also a global solution.*

PROOF. Suppose for contradiction that there is a feasible point $\bar{x}$ with $f(\bar{x}) < f(x^*)$. Then by convexity of the feasible region, we have that $\lambda x^* + (1 - \lambda)\bar{x}$ is also feasible, for any $\lambda \in [0, 1]$. By convexity of f, we have that

$$f(\lambda x^* + (1 - \lambda)\bar{x}) \leq \lambda f(x^*) + (1 - \lambda) f(\bar{x}) < f(x^*)$$

for all $\lambda \in [0, 1)$. Since any neighborhood of x^* will contain one of the points along this line, we cannot identify a neighborhood that satisfies the definition of a local solution at x^*. □

Convex programming can be analyzed by the tools of convex analysis, an extremely elegant area of mathematics which is the subject of the classic monograph by Rockafellar [**Roc70**]. This theory uses a geometric approach that complements the algebraic approach usually used in the development of algorithms. The recent book by Bertsekas, Nedić, and Ozdaglar [**BNO03**] explores further the connections between convex analysis and optimization. Convex analysis has given rise to nonsmooth analysis [**Cla83**] and variational analysis [**RW98**], areas that are providing the tools for studying and solving an ever-wider class of optimization problems.

3. Nonlinear Programming: Algorithms

Algorithms for nonlinear programming are usually iterative in nature, in that they generate a sequence of estimates $\{x^k\}_{k=0,1,2,\ldots}$ of the solution x^* and possibly a sequence of estimates of the optimal Lagrange multipliers λ^* and γ^* as well. The process of generating each iterate x^k may itself be iterative, so that we sometimes speak of "inner" and "outer" iterations. Each iterate x^k is usually obtained by solving a "subproblem," which is often a simplified version of (2.1) or some other type of optimization, complementarity, or linear algebra problem whose solution approximates a solution of the original problem (2.1).

Our assumption of smoothness of the objective and constraints is essential to the effectiveness of the algorithms discussed in this paper. The subproblems often are constructed from linear or quadratic approximations of f, c, and d, and the validity of these approximations depends on Taylor's theorem and on the smoothness properties.

One important issue in designing algorithms for (2.1) concerns the decision about whether or not to take the step to a new point proposed by the subproblem. The issue is to judge the "merit" of the new point relative to the current point (and possibly also to earlier points encountered by the algorithm). Design of a measure of merit is not straightforward, given that we wish both to attain feasibility and to minimize the objective over the feasible region. When no constraint is present, the issue becomes simple—the objective function itself is a measure of merit. Indeed, certain algorithms for (2.1) ensure that all iterates $\{x^k\}$ remain feasible, and for such algorithms, f can indeed be used as a measure of merit. Other algorithms define a merit function to be a weighted sum of f and a measure of the constraint violation, thus "trading off" the goals of attaining feasibility and minimizing the objective. A newer class of methods known as "filter methods" does not try to perform this tradeoff. Instead, these methods maintain a "filter" consisting of a collection of iterates generated at previous iterations, and accepts a new point if there is no point currently in the filter that has both a better function value and a smaller constraint violation than the new point.

The analysis of convergence of nonlinear programming algorithms yields results of two major types. The first class includes global convergence results, which are typically of the form that "the sequence generated by the algorithm has a limit point that is stationary for the problem (2.1)" or "all limit points are stationary." The second class are local convergence results. A typical result of this type is as follows: "When started sufficiently close to a solution x^* satisfying second-order sufficient conditions, the algorithm converges superlinearly to x^*."

We now outline several classes of algorithms for solving nonlinear programming problems, indicating for each class the availability of software tools based on that class. Our list of algorithms is far from exhaustive; we focus on those that have proved to be the most practical or influential.

3.1. Newton's Method. In subsequent sections we make occasional references to Newton's method, a technique that is fundamental to optimization over continuous variables. We describe this technique briefly here.

For the unconstrained minimization problem

$$\min_x f(x)$$

for a smooth (twice continuously differentiable) function $f : \mathbf{R}^n \to \mathbf{R}$, Newton's method obtains a step from the current iterate x by forming a quadratic Taylor-series approximation to f around x and minimizing this approximation. That is, the step Δx is chosen to satisfy

$$\min_{\Delta x} f(x) + \nabla f(x)^T \Delta x + \frac{1}{2} \Delta x^T \nabla^2 f(x) \Delta x. \tag{3.1}$$

We then choose the next iterate to be $x + \Delta x$. Obviously the subproblem (3.1) has a finite solution only if $\nabla^2 f(x)$ is positive semidefinite, and this condition may only hold in the neighborhood of a local minimizer satisfying second-order sufficient conditions. However, when started in such a neighborhood, Newton's method converges rapidly, so it is the foundation of most approaches for minimizing smooth functions. Various modifications are needed to make Newton's method a practical, robust algorithm.

Another variant of Newton's method is used to find solutions of nonlinear algebraic systems of the form

$$F(x) = 0,$$

where $F : \mathbf{R}^n \to \mathbf{R}^n$ is a smooth (continuously differentiable) vector function. Defining $\nabla F(x)$ to be the $n \times n$ matrix of first partial derivatives of the components of F, we obtain a step Δx by solving the following linear system of equations:

$$F(x) + \nabla F(x) \Delta x = 0. \tag{3.2}$$

Again, this technique converges rapidly to a solution x^* at which $\nabla F(x^*)$ is nonsingular when the starting point is sufficiently close to x^*. As before, modifications are needed to ensure robust behavior (convergence from remote starting points, convergence to solutions at which $\nabla F(x^*)$ is singular, and so on).

Both these variants of Newton's method require knowledge of derivatives. Formulas for these derivatives can be supplied by the user, or automatic differentiation tools can be used (see Griewank [**Gri00**]) to obtain exact derivatives. Alternatively, numerical differentiation can be used to compute approximate derivatives, based on the values of functions or lower-order derivatives.

Another class of methods avoids computing exact or approximate derivative matrices. In the case of minimization, these methods use instead an approximate Hessian H in place of $\nabla^2 f(x)$ in (3.1), updating H after each iteration in response to experience gained over the latest steps. These *quasi-Newton methods*, which first appeared in the early 1960s, have proved to be very successful in practice. Variants have been developed for constrained optimization and for large-scale problems that require implicit storage of H.

3.2. Barrier Algorithm. Barrier algorithms were among the first class of algorithms proposed for solving nonlinear programs. We outline the most popular of these methods—the logarithmic barrier method—for the case in which there are no equality constraints (d vacuous) and for which there is at least one x with $c(x) > 0$ (strict inequality for each component of c).

The logarithmic barrier function for the problem

$$\min_x f(x) \qquad \text{subject to } c(x) \geq 0 \tag{3.3}$$

is defined as follows:

$$P(x;\mu) \stackrel{\text{def}}{=} f(x) - \mu \sum_{i=1}^{m} \log c_i(x), \tag{3.4}$$

where μ is a positive scalar known as the barrier parameter. The domain of this function is the set Ω_0 defined by

$$\Omega_0 \stackrel{\text{def}}{=} \{x \mid c(x) > 0\}.$$

This function approaches ∞ as x approaches any point $\bar{x}$ with $c_i(\bar{x}) = 0$ for some i. However, when x is "not too close" to the boundary of the feasible region and μ is small, $P(x;\mu)$ is quite close to $f(x)$. As $\mu \downarrow 0$, the region within which $P(x;\mu)$ and $f(x)$ diverge significantly shrinks to a narrow layer at the boundary of the feasible set.

The algorithmic usefulness of the logarithmic barrier function derives from the following result, which shows that local minimizers of $P(\cdot;\mu_k)$ approach local solutions of the problem (3.3) under certain assumptions. This result makes use of the following definition. Its proof can be found in the classic book on the subject by Fiacco and McCormick [**FM68**].

DEFINITION 3.1. A nonempty set $M^* \subset M$ is called an *isolated set of* M if there exists a closed set E such that $M^* \subset \text{interior}(E)$ and in addition $x \in E \setminus M^*$ implies $x \notin M$.

THEOREM 3.2. *Suppose that f and c are continuous, that the set of local solutions A^* corresponding to a certain value f^* of the objective is a nonempty, compact, and an isolated set of $\{x \in \Omega \mid f(x) = f^*\}$, and that at least one point in A^* is in the closure of Ω_0. Let $\{\mu_k\}$ be a strictly decreasing sequence of positive scalars. Then there exists a compact set S such that $A^* \subset \text{interior}(S)$ and the minimizer of $P(\cdot;\mu_k)$ in $\Omega_0 \cap \text{interior}(S)$ exists for all k sufficiently large, and every limit point of every subsequence $\{x^k\}$ of these minimizing points is in A^*. Moreover, we have that $\{f(x^k)\}$ is a monotonically decreasing sequence, and $f(x^k) \downarrow f^*$.*

The review paper of M. Wright [**Wri92**] contains other interesting theoretical results about the log barrier approach.

Theorem 3.2 suggests an algorithm in which we find approximate minimizers of $P(\cdot;\mu_k)$ over Ω_0 for a decreasing sequence $\{\mu_k\}$, using information about the approximate minimizers for $\mu_0, \mu_1, \ldots, \mu_k$ to identify a starting point for the minimization of $P(\cdot;\mu_{k+1})$. Since each $P(\cdot;\mu_k)$ is as smooth as are the functions f and c over Ω_0, a suitably safeguarded scheme for unconstrained minimization can in principle be used.

This approach was used in the 1960s and 1970s, but its popularity declined when a key disadvantage was identified. When μ_k is small, the Hessian $\nabla^2_{xx} P(\cdot;\mu_k)$ is ill-conditioned in the vicinity of any minimizer of $P(\cdot;\mu_k)$. This property leads to a small radius of convergence of the minimization process for $P(\cdot;\mu_k)$; the starting guess has to be quite close to the minimizer for the optimization process to converge reliably. Additionally, if a Newton method is used to calculate the steps of the unconstrained minimization, numerical error in the linear algebra computations make the computed step less and less accurate. Recent research has shown that these difficulties can be ameliorated to some extent by the use of extrapolation schemes or reformulation techniques to define starting points close to the minimizer of $P(\cdot;\mu_k)$. Careful analysis has also revealed that the numerical errors arising from the ill-conditioning tend to be limited to a subspace parallel to the constraint boundaries, and that the overall convergence need not be affected dramatically by these errors.

When equality constraints $d(x) = 0$ are also present, they can be handled by including a quadratic penalty term (see Section 3.3) or by enforcing them explicitly, so that each subproblem becomes the following equality constrained optimization problem:

$$\min_x P(x;\mu) \stackrel{\text{def}}{=} f(x) - \mu \sum_{i=1}^{m} \log c_i(x), \quad \text{subject to } d(x) = 0. \tag{3.5}$$

There currently appears to be no production software code based purely on the log barrier approach. The method's importance lies largely in its connection to the interior-point methods rediscovered in the 1980s and 1990s (discussed below in Section 3.5) and its relevance to successful algorithms for linear programming and semidefinite programming.

3.3. Penalty and Augmented Lagrangian Methods. Like barrier methods, penalty methods define a parametrized function that combines the objective and constraints and which can be minimized using techniques for unconstrained optimization. Unlike in barrier methods, the domain of penalty functions extends beyond the feasible set, and their value on the feasible set equals the value of the objective. Violations of the constraints are penalized but not forbidden. Solutions to (2.1) are found by minimizing the penalty function for an increasing or decreasing sequence of values of the penalty parameter.

Perhaps the simplest penalty function is the quadratic penalty function, defined as follows:

$$Q(x;\mu) = f(x) + (2\mu)^{-1}\|\min(0, c(x))\|_2^2 + (2\mu)^{-1}\|d(x)\|_2^2, \tag{3.6}$$

where $\min(0, c(x))$ is the vector whose ith component is $\min(0, c_i(x))$, and where $\mu > 0$ is a penalty parameter. Even when f, c, and d are smooth, this function has a discontinuity in its second derivative (arising from the min function in the second term.) When there are no inequality constraints (c vacuous), the function $Q(\cdot;\mu)$ is as smooth as are f and d.

The minimizer of $Q(x;\mu)$ for any given $\mu > 0$ will generally be an infeasible point. As $\mu \to 0$, however, the penalty for constraint violation becomes more severe, and we would expect the violation to decrease. In fact, it can be shown that the sequence of minimizers of $Q(\cdot;\mu)$ approaches a local solution of (2.1) in much the same fashion as do the minimizers of the logarithmic barrier function (see Theorem 3.2). At the minimizer of $Q(\cdot;\mu)$, we have $\nabla Q(x;\mu) = 0$, that is,

$$0 = \nabla Q(x;\mu) = \nabla f(x) + \nabla c(x)[\min(0, c(x))/\mu] + \nabla d(x)[d(x)/\mu].$$

The similarity between this equation and (2.4a) (they are both linear combinations of the gradients ∇f, $\nabla c_i(x)$, $i = 1, 2, \ldots, m$, and $\nabla d_i(x)$, $i = 1, 2, \ldots, p$) suggests that in the vicinity of a local solution x^* satisfying certain assumptions, we have $x \approx x^*$, $\lambda^* \approx -\min(0, c(x))/\mu$, and $\gamma^* \approx -d(x)/\mu)$. These relationships suggest however that as in the barrier approach, there is a systematic offset between the minimizer of $Q(\cdot;\mu)$ and the solution x^*—we have $d(x) \approx -\mu\gamma^*$ and $\min(0, c(x)) \approx -\mu\lambda^*$—that can be resolved only by driving μ to zero. Unfortunately, as in the log barrier method, small values of μ degrade the conditioning of the minimization problem (3.6) and make it difficult to solve numerically.

A modification of the quadratic penalty approach avoids the "offset" problem by making use of the min-max formulation (2.9) and the *proximal point* algorithm [**Roc76**]. Supposing that prior estimates λ^k and γ^k of the optimal Lagrange multipliers are known, we modify the problem (2.9) as follows:

$$\min_x \max_{\lambda \geq 0, \gamma} \mathcal{L}(x, \lambda, \gamma) - (\mu_k/2)\|\lambda - \lambda^k\|_2^2 - (\mu_k/2)\|\gamma - \gamma^k\|^2, \tag{3.7}$$

where μ_k is a small positive parameter. The additional terms serve to penalize any step away from the current multiplier estimates (λ^k, γ^k). They also make it possible to find the maximizing values of $\lambda \geq 0$ and γ explicitly in (3.7), since the function being maximized is a convex quadratic in both these variables. By setting derivatives with respect to λ_i and γ_i of the function in (3.7) to zero and taking the constraints $\lambda_i \geq 0$, $i = 1, 2, \ldots, m$ into account, we obtain

$$\lambda_i = \max(0, \lambda_i^k - c_i(x)/\mu_k), \quad i = 1, 2, \ldots, m, \tag{3.8a}$$

$$\gamma_i = \gamma_i^k - d_i(x)/\mu_k, \quad i = 1, 2, \ldots, p. \tag{3.8b}$$

By substituting these values into (3.7) and rearranging, we obtain

$$\begin{aligned} \min_x \; & f(x) - (\gamma^k)^T d(x) + (2\mu_k)^{-1}\|d(x)\|_2^2 \\ & + (\mu_k/2)\|\max(\lambda^k - c(x)/\mu_k, 0)\|_2^2 - (\mu_k/2)\|\lambda^k\|_2^2. \end{aligned} \tag{3.9}$$

The objective function in (3.9) is known as the *augmented Lagrangian.*

An algorithm based on (3.9) makes a succession of estimates λ^k and γ^k, and then solves (3.9) for a given value of the penalty parameter μ_k to obtain an estimate of the solution x^*. The *method of multipliers* uses the formulas (3.8) to update the Lagrange multipliers; that is, given an approximate minimizer x^k of (3.9), it sets

$$\lambda_i^{k+1} = \max(0, \lambda_i^k - c_i(x^k)/\mu_k), \quad i = 1, 2, \ldots, m, \tag{3.10a}$$

$$\gamma_i^{k+1} = \gamma_i^k - d_i(x^k)/\mu_k, \quad i = 1, 2, \ldots, p. \tag{3.10b}$$

It then selects μ_{k+1} by some heuristic, and solves (3.9) (with k replaced by $k+1$) to obtain the next estimate x^{k+1} of x^*.

This approach has the advantage that it is not generally necessary to decrease μ_k to zero in order for x^k to be a good approximation to x^*. In fact, under standard assumptions on the local minimizer x^*, it is not difficult to show that if the Lagrange multipliers are exact in (3.9) (that is, $\lambda^k = \lambda^*$ and $\gamma^k = \gamma^*$), then x^* is a local minimizer of the unconstrained problem (3.9) for all μ_k sufficiently small. This result suggests that it is not necessary to drive μ_k to zero to obtain an excellent approximation to x^*, so the unconstrained subproblem (3.9) need not become excessively ill conditioned.

The method of multipliers was proposed originally by Powell [**Pow69**] and Hestenes [**Hes69**] for the equality-constrained case, and extended to the case of inequality constraints by Rockafellar [**Roc71, Roc73**]. The methods are described in the comprehensive monograph by Bertsekas [**Ber82**]. Conn, Gould, and Toint [**CGT91, CGT92**] based their successful code Lancelot on this approach.

An alternative is to use a *nonsmooth* penalty function. The best-known function of this type is the ℓ_1 *penalty function*, which is defined as follows:

$$f(x) + \mu^{-1}\|\min(0, c(x))\|_1 + \mu^{-1}\|d(x)\|_1, \tag{3.11}$$

It can be shown that under standard assumptions, a local solution of the problem (2.1) is a local minimizer of (3.11), for all μ sufficiently small. Unfortunately, the function (3.11) is nonsmooth; in fact, its minimizer generally lies at a point at which the first derivative is not defined. The nonsmoothness enters the function in a highly structured way, however, that makes it possible to use specialized techniques for minimizing (3.11) rather than resorting to general methods for continuous but nonsmooth functions.

3.4. Sequential Quadratic Programming. The sequential quadratic programming (SQP) approach generates steps via a quadratic programming approximation to (2.1), based on Taylor series expansions of the objective and constraint functions around the current iterate x^k. In the most basic form of SQP, the subproblem is as follows:

$$\min_{\Delta x} \nabla f(x^k)^T \Delta x + \tfrac{1}{2}\Delta x^T H_k \Delta x \quad \text{subject to} \tag{3.12a}$$

$$c(x^k) + \nabla c(x^k)^T \Delta x \geq 0, \tag{3.12b}$$

$$d(x^k) + \nabla d(x^k)^T \Delta x = 0, \tag{3.12c}$$

where H_k is a symmetric matrix. When H_k is positive definite, the quadratic program (3.12) is convex and tractable; otherwise it is nonconvex and the subproblem may be unbounded or have multiple local minima. The Lagrange multipliers for the constraints (3.12b), (3.12c) can be used as estimates for the Lagrange multipliers λ^* and γ^* in the original problem (2.1)

In a sense, the "ideal" choice for H_k is the Hessian of the Lagrangian $\nabla^2_{xx}\mathcal{L}(x^k, \lambda^k, \gamma^k)$ evaluated at x^k and the current estimate of the Lagrange multipliers λ^k and γ^k. Note that $\nabla^2_{xx}\mathcal{L}(x^k, \lambda^k, \gamma^k)$ incorporates curvature information from not only the objective but also the constraint functions. For this choice, under suitable assumptions on the local minimizer x^* and defining Δx^k to be the local solution of (3.12) with smallest norm, the sequence generated by setting $x^{k+1} \leftarrow x^k + \Delta x^k$ and $(\lambda^{k+1}, \gamma^{k+1})$ to the Lagrange multipliers for (3.12b), (3.12c) converges quadratically to $(x^*, \lambda^*, \gamma^*)$. Variants of this approach choose H_k to be a *quasi-Newton* approximation to $\nabla^2_{xx}\mathcal{L}(x^k, \lambda^k, \gamma^k)$, obtained from changes to the gradient $\nabla_x \mathcal{L}(\cdot, \lambda^k, \gamma^k)$ over the current step in a similar way to quasi-Newton methods for unconstrained optimization.

The SQP approach appears to originate with Wilson [**Wil63**]. A quasi-Newton variant is described by Powell [**Pow77**]. Other quasi-Newton methods maintain an approximation to the projection of $\nabla^2_{xx}\mathcal{L}(x^k, \lambda^k, \gamma^k)$ onto the null space of the active constraint gradients, which corresponds to the set $\mathcal{C}$ of (2.7) when there are no constraints $i \in \mathcal{A}(x^*)$ with $\lambda^*_i = 0$. As can be seen from Theorem 2.4, this projected Hessian can be expected to be positive semidefinite in a neighborhood of a solution, so the use of a positive definite quasi-Newton approximation (as is done in the case of unconstrained optimization) is appropriate. Since in many applications the projected Hessian is a much smaller matrix than the full Hessian, this approach can lead to practical and efficient algorithms.

As described so far, the SQP approach will function well only in a neighborhood of the solution x^*. A number of enhancements are needed to turn this approach into a practical algorithm. Various extensions of the line-search and trust-region ideas from unconstrained optimization have been investigated, but these simple ideas are made considerably more complicated by the presence of constraints. If, for instance, we add the trust-region constraint $\|\Delta x\| \leq \rho$ to the SQP subproblem

(3.12) (thus limiting the step length to a region in which we believe the Taylor series approximations to the objective and constraint functions are valid), we will find that the subproblem becomes infeasible if x^k is infeasible for (2.1) and ρ is sufficiently small. Modifications of the trust-region approach have been suggested that split the SQP step into two components—one that decreases the infeasibility and another that improves the objective (see Omojokun [**Omo89**])—but these have not been implemented in the context of a production SQP code.

Another issue concerns which merit function should be used to decide on acceptability of a candidate step. This issue arises also in line-search algorithms, which try steps of the form $\alpha\Delta x$ for a sequence of decreasing values of α until one of the candidate steps $x + \alpha\Delta x$ is found to be acceptable. Han [**Han77**] showed that the use of the nonsmooth penalty function (3.11) as a merit function yielded an algorithm with good global convergence properties. However, Maratos [**Mar78**] showed that the local convergence properties can be poor. Arbitrarily close to the solution, the merit function may actually increase for the full step generated by (3.12), even though the step is a good one in that it yields an iterate x^{k+1} much closer to the solution than x^k. Hence, good steps may be rejected and the algorithm may converge only at a linear rate. This "Maratos effect" can be avoided by the use of a second-order correction, which is a modification of the step Δx obtained by solving an auxiliary quadratic program.

In all nonlinear optimization algorithms, the two main components of the computational cost are the cost of the evaluations of the functions and their derivatives (where needed) and the cost of forming and solving the subproblems. In SQP, the cost of solving the subproblem is often nontrivial, so this approach tends to be best suited to problems for which the function and derivative evaluations are relatively expensive. The code SNOPT [**GMS02**] is a sophisticated modern implementation of the SQP approach.

3.5. Interior-Point Methods. Modern interior-point methods for solving (2.1) can be motivated in two ways. The first is as approximate Newton methods applied to the KKT conditions (2.4), following similar strategies in the choice of search direction as the most successful variants of interior-point algorithms for linear programming. The second derives from the optimality conditions for the barrier subproblem (3.5), with Lagrange multiplier estimates introduced explicitly as variables.

For the first motivation, we introduce slack variables s for the inequality constraints and write the conditions (2.4) as follows:

$$\nabla f(x) - \nabla c(x)\lambda - \nabla d(x)\gamma = 0, \tag{3.13a}$$

$$d(x) = 0, \tag{3.13b}$$

$$c(x) - s = 0, \tag{3.13c}$$

$$\lambda \geq 0, \tag{3.13d}$$

$$s \geq 0, \tag{3.13e}$$

$$\lambda^T s = 0. \tag{3.13f}$$

These conditions can be formulated as the bounds $\lambda \geq 0$, $s \geq 0$ together with the following system of nonlinear equations:

$$F(x, s, \lambda, \gamma) \stackrel{\text{def}}{=} \begin{bmatrix} \nabla f(x) - \nabla c(x)\lambda - \nabla d(x)\gamma \\ d(x) \\ c(x) - s \\ \Lambda S e \end{bmatrix} = 0, \tag{3.14}$$

where $\Lambda = \text{diag}(\lambda_1, \lambda_2, \ldots, \lambda_m)$, $S = \text{diag}(s_1, s_2, \ldots, s_m)$, and $e = (1, 1, \ldots, 1)^T$. Note that the system (3.14) has the same number of variables as equations—$n + 2m + p$—so methods of the Newton type can be applied. The search direction is modified and the line search step size is chosen so that the s and λ components of all iterates $(x^k, s^k, \lambda^k, \gamma^k)$ remain strictly positive for all $k = 0, 1, 2, \ldots$. The latter requirement gives rise to the term "interior-point." (This term is however somewhat misleading, since the iterates $(x^k, s^k, \lambda^k, \gamma^k)$ are not necessarily in the interior of the feasible region; in fact they are often infeasible because the equality constraints (3.13a), (3.13b), (3.13c) are usually violated.) The quantity $\mu = s^T\lambda/m$ can be used to measure the violation of the condition (3.13f); it is sometimes known as the "complementarity measure." The search directions $(\Delta x, \Delta s, \Delta\lambda, \Delta\gamma)$ at a point (x, s, λ, γ) are often obtained by solving the following system of equations, for some parameter $\sigma \in [0, 1]$:

$$\nabla F(x, s, \lambda, \gamma) \begin{bmatrix} \Delta x \\ \Delta s \\ \Delta\lambda \\ \Delta\gamma \end{bmatrix} = -F(x, s, \lambda, \gamma) + \begin{bmatrix} 0 \\ 0 \\ 0 \\ \sigma\mu e \end{bmatrix}. \tag{3.15}$$

If $\sigma = 0$, this is simply a Newton step for the system $F(x, s, \lambda, \gamma) = 0$. The term $\sigma\mu e$ serves to "center" the step, making it a Newton step toward the point satisfying $F(x, s, \lambda, \gamma) - (0, 0, 0, \sigma\mu e) = 0$, and allowing a longer step to be taken along this direction without violating positivity of the s and λ components.

For the second motivation, we write the KKT conditions for the problem (3.5) as follows:

$$\nabla f(x) - \nabla c(x)[\mu/c(x)] - \nabla d(x)\gamma = 0, \tag{3.16a}$$

$$d(x) = 0, \tag{3.16b}$$

$$c(x) > 0, \tag{3.16c}$$

where $\mu/c(x)$ denotes the vector whose components are $\mu/c_i(x)$. Denoting this vector by λ, and introducing slacks $s = c(x)$ into the formulation, we see that (3.16) is identical to the system

$$F(x, s, \lambda, \gamma) - (0, 0, 0, \mu e) = 0.$$

It can be shown that the Δx and $\Delta\gamma$ components of the step obtained from (3.15) with $\sigma = 1$ are identical to the step obtained by applying Newton's method to the two equations (3.16a), (3.16b) when $\lambda = \mu/c(x)$ and $s = c(x)$. However, the essence of the primal-dual approach—the feature that distinguishes it from the logarithmic-barrier approach of Section 3.2—is to treat λ and s as independent variables, rather than to derive their values from x and γ.

For the special case in which (2.1) is a convex programming problem, it is not difficult to design algorithms based on steps of the form (3.15) that converge globally and also have rapid local convergence, under standard assumptions (see for example

Wright and Ralph [**WR96**]). Such algorithms usually choose the steplength in such a way that the complementarity measure μ decreases significantly at each iteration, while the λ and s vectors stay positive, and violations of the equality constraints (3.13a), (3.13b), and (3.13c) remain bounded by some multiple of μ. They are similar to path-following primal-dual interior-point algorithms for linear programming.

For nonconvex problems, however, more complex strategies are needed to ensure global convergence. One issue is that the top left block in the matrix ∇F in (3.15)—which is simply the Hessian of the Lagrangian (2.5)—may be singular or have negative curvature, resulting in possible singularity of ∇F. Some algorithms (such as LOQO [**SBV02**]) simply add a sufficiently large multiple of the identity to this matrix to ensure that the step is well defined. The KNITRO-DIRECT approach [**WMNO03**] falls back on an alternative means for calculating the step when the eigenvalue distribution of ∇F is one that may produce an inappropriate step. (Specifically, it uses a conjugate-gradient approach to calculate the projection of the $(\Delta x, \Delta s)$ step component into the null space of the constraint Jacobian.)

Another important feature that distinguishes the nonconvex case from the convex and linear cases was pointed out by Wächter and Biegler [**WB00**]. If the steps $(\Delta x, \Delta s, \Delta\lambda, \Delta\gamma)$ satisfy the linearizations of (3.13b) and (3.13c), that is,

$$d(x) + \nabla d(x)^T \Delta x = 0, \qquad c(x) + \nabla c(x)^T \Delta z - s - \Delta s = 0, \tag{3.17}$$

and if the step length is chosen to maintain positivity of λ and s, then the algorithm may converge to an infeasible point. Since the step obtained from (3.15) will certainly satisfy these conditions, it cannot form the basis of a reliable algorithm. A number of approaches have been proposed to ameliorate this problem. One such approach is to require only a fraction of the best attainable improvement in feasibility of the constraints (3.13b), (3.13c) at each iteration (Byrd, Hribar, and Nocedal [**BHN99**]). A second approach is to use "feasibility restoration" steps at certain points in the algorithm (Wächter [**Wae02**]).

The choice of step length along the direction computed from (3.15) is governed first by the requirement that the s and λ components of all iterates remain strictly positive, and second by the use of a merit function or "filter" approach. Wächter [**Wae02**] uses a filter approach in which each step is required to give an improvement either in the violation of the constraints (3.13b), (3.13c) or in the primal barrier function for the current value of μ. The KNITRO-Direct code (Waltz et al. [**WMNO03**]) uses a linear combination of the primal barrier function and the norm of the constraint violations.

Interior-point software for nonlinear programming has taken some years to attain the success of the same type of software for linear programming. However, codes such as KNITRO, LOQO, and IPOPT have appeared recently and have proved to be highly competitive with the best existing codes, and superior on many classes of problems. They hold a particular advantage over SQP codes in cases where the function and derivative evaluations are not particularly expensive. The cost of solving the subproblem is usually considerably cheaper than for SQP, so they can produce an improvement in runtime over SQP codes even when more outer iterations are required for convergence.

4. Semidefinite Programming: Origins and Optimality Theory

In semidefinite programming, the variable is not a vector of real numbers but rather a symmetric matrix that is constrained to be positive semidefinite. Denoting the space of symmetric $n \times n$ matrices as $S\mathbb{R}^{n\times n}$, we define the standard form of this problem as follows:

$$(4.1) \qquad \min_{X\in S\mathbb{R}^{n\times n}} C \bullet X \quad \text{s.t.} \quad A_i \bullet X = b_i, \; i = 1, 2, \ldots, m; \qquad X \succeq 0,$$

where we write $U \succeq V$ for any two matrices U and V in $S\mathbb{R}^{n\times n}$ to indicate that $U - V$ is positive semidefinite, and $U \bullet V$ denotes $\sum_{i,j=1}^{n} U_{ij}V_{ij}$ or, equivalently, trace(UV). We require here that C and A_i are all symmetric $n \times n$ matrices.

Semidefinite programming admits an elegant theory of duality, which we discuss in Section 4.1. For the moment, we define the dual problem for (4.1), a problem that is made up from the same data as (4.1) (arranged differently) and closely related to it:

$$(4.2) \qquad \max_{y\in\mathbb{R}^m, S\in S\mathbb{R}^{n\times n}} b^T y \quad \text{s.t.} \quad \sum_{i=1}^{m} A_i y_i + S = C, \qquad S \succeq 0.$$

Note that linear programming is a special case of semidefinite programming. If all the data objects C, A_i, $i = 1, 2, \ldots, m$ are diagonal, then (4.1) is equivalent to the linear programming problem in standard form, where the vector of variables is formed by the diagonal of X.

Semidefinite programming problems have a wealth of applications to such diverse areas as control [**VB96**], electronic structure calculations [**ZBF⁺03**], and signal processing; and to other areas of optimization such as robust optimization, combinatorial optimization, and polynomial programming (global optimization involving polynomials).

A classic problem in semidefinite programming is minimizing the maximum eigenvalue of an affine matrix function:

$$(4.3) \qquad \min_{z\in\mathbb{R}^n} \lambda_{\max}\left(A_0 + \sum_{i=1}^{n} A_i z_i\right),$$

which can be reformulated as follows:

$$\max_{\alpha, z} -\alpha \quad \text{s.t.} \quad \alpha I - \left(A_0 + \sum_{i=1}^{n} A_i z_i\right) - S = 0, \quad S \succeq 0,$$

which clearly has the form (4.2) when we define $y = (\alpha, z)$.

A second application is from control theory. Lyapunov theory has that a linear differential equation

$$\dot{x}(t) = Ax(t), \qquad x(0) = x_0$$

(where $x(t) \in \mathbb{R}^n$) is asymptotically stable (all trajectories converge to zero) if and only if there is a positive definite matrix P such that $A^T P + PA \prec 0$ (that is, $A^T P + PA$ is negative definite). Similarly, all trajectories remain bounded if and only if there is a positive definite matrix P such that $A^T P + PA \preceq 0$. This theory can be extended to the case of a differential inclusion

$$\dot{x}(t) \in \text{conv}\{A_1, A_2, \ldots, A_m\}x(t), \qquad x(0) = x_0,$$

by seeking a positive definite P such that

$$A_i^T P + P A_i \preceq 0, \qquad \text{for all } i = 1, 2, \dots, m.$$

Since these constraints are homogeneous, we can replace the requirement $P \succ 0$ by $P \succeq I$, and we can turn the problem into an optimization problem by seeking the P with smallest maximum eigenvalue. We obtain

$$\max_{\eta, P} -\eta \quad \text{s.t.} \quad A_i^T P + P A_i \preceq 0, \qquad \text{for all } i = 1, 2, \dots, m; \quad \eta I \succeq P \succeq I,$$

which can be expressed as a semidefinite program.

Semidefinite programming problems cannot be formulated easily as nonlinear programs in which the elements of the matrix X are the variables; the semidefiniteness condition $X \succeq 0$ translates to an unpleasant nonlinear constraint. Special cases (especially (4.3)) have been solved successfully with algorithms for nonsmooth nonlinear programming (see Overton and Womersley [**OW93**] and Oustry [**Ous00**]). However the bulk of research effort in this area since 1992 has been on methods based on the interior-point approach. The work of Alizadeh [**Ali91, Ali92**] and the monograph of Nesterov and Nemirovski [**NN94**] were instrumental in founding this research area.

There are several excellent review papers and books on semidefinite programming; we mention just a few here, and refer the reader to their bibliographies for additional material. Todd [**Tod01**] sketches the history of the area and gives an authoritative survey of both applications and algorithmic developments. (We draw extensively on this paper in these notes.) The handbook edited by Wolkowicz et al. [**WSV00**] contains contributions on many aspects of the area. The paper of Vandenberghe and Boyd [**VB96**] and the book of Ben-Tal and Nemirovski [**BN01**] contain a wealth of applications. Nesterov and Nemirovski's monograph [**NN94**] developed the theory of self-concordant barriers for optimization over cones (of which semidefinite programming is an important special case), which opened the door to powerful classes of interior-point algorithms. Nesterov and Todd [**NT97, NT98**] studied self-scaled cones, which give rise to primal-dual interior-point algorithms. Derivation of interior-point algorithms from Euclidean Jordan algebras was proposed by Faybusovich [**Fay97a, Fay97b**].

4.1. Duality Theory. Semidefinite programming admits a duality theory that is similar to but considerably more complex than that of linear programming. The weak duality result is trivial.

THEOREM 4.1 (Weak Duality). *If X is feasible for (4.1) and (y, S) is feasible for (4.2), then*

$$C \bullet X - b^T y = X \bullet S \geq 0.$$

PROOF. We have by substitution from the constraints in (4.1) and (4.2) that

$$C \bullet X - b^T y = \left(\sum_{i=1}^m y_i A_i + S \right) \bullet X - b^T y = \sum_{i=1}^m y_i (A_i \bullet X) + S \bullet X - b^T y = S \bullet X = X \bullet S.$$

Since X is positive semidefinite, it has a square root $X^{1/2}$. Since $\text{trace}(PQ) = \text{trace}(QP)$ for any symmetric Q and P, we have

$$X \bullet S = \text{trace}(XS) = \text{trace}(X^{1/2} X^{1/2} S) = \text{trace}(X^{1/2} S X^{1/2}).$$

Since the trace of a symmetric matrix is the sum of its eigenvalues, we have

$$X \bullet S = \sum_{j=1}^{n} \lambda_j (X^{1/2} S X^{1/2}) \geq 0,$$

where λ_j denotes the jth eigenvalue. The final inequality follows from the fact that $X^{1/2}SX^{1/2}$ is positive semidefinite. □

The strong duality result is different from linear programming. Examples are given in [**Tod01**, Section 4] and [**VB96**] in which both (4.1) and (4.2) are feasible, but where one of the two fails to attain its minimum value, or where both attain their minimum value but the gap $C \bullet X - b^T y$ remains positive. However, strong duality results can be proved for cases in which the feasible regions of (4.1) and (4.2) have nonempty interiors. Proofs of the following results can be found in Todd [**Tod01**, Section 4].

THEOREM 4.2. *Suppose that (4.1) and (4.2) are feasible and that the feasible region for (4.2) contains an element (y, S) with positive definite S. Then the set of matrices X that solve the problem (4.1) is nonempty and compact, and the optimal values of (4.1) and (4.2) are identical.*

Similarly, if (4.1) and (4.2) are feasible and the feasible region for (4.1) contains a positive definite matrix X, then the set of pairs (y, S) that solve (4.2) is nonempty and compact, and the optimal values of (4.1) and (4.2) are identical.

COROLLARY 4.3. *Suppose that (4.1) and (4.2) are feasible and that there exist positive definite matrices X and S such that X is feasible for (4.1) and (y, S) is feasible for (4.2), for some y. Then both (4.1) and (4.2) have nonempty compact sets of solutions, and the optimal values are identical.*

4.2. Barrier Functions. Nesterov and Nemirovski [**NN94**] laid an important foundation for semidefinite programming algorithms when they defined *self-concordant barrier functions.* Given a convex cone K in a finite-dimensional real vector space with nonempty interior that is pointed (containing no lines), a self-concordant barrier function is a function F : interior$(K) \to \mathbb{R}$ with the following properties:

- F is convex and three times differentiable;
- $F(x) \to \infty$ as x approaches the boundary of K;
- For all $x \in$ interior(K) and all h, we have $|F'''(x)hhh| \leq 2(F''(x)hh)^{3/2}$;
- There is a constant $\vartheta > 0$ such that for all $x \in$ interior(K) we have

$$\langle F'(x), [F''(x)]^{-1} F'(x) \rangle \leq \vartheta.$$

The second of these properties is common to all barrier functions. The third suggests that F is "not too non-quadratic," in that its third directional derivatives are bounded in terms of its second directional derivatives. The quantity appearing in the fourth property is the weighted norm of the Newton step $[F''(x)]^{-1}F'(x)$, where the weight is $F''(x)$ itself.

Nesterov and Nemirovski show that such functions exist for any pointed convex cone K with nonempty interior. They are difficult to calculate in general, but can be identified explicitly in three important special cases:

- $K = \{x \in \mathbb{R}^n \mid x \geq 0\}$:

$$F(x) = -\sum_{i=1}^{n} \log x_i, \qquad \text{with } \vartheta = n; \tag{4.4}$$

- $K = \{X \in S\mathbb{R}^{n \times n} \mid X \succeq 0\}$:

$$F(X) = -\log \det X, \qquad \text{with } \vartheta = n; \tag{4.5}$$

- $K = \{(x, t) \in \mathbb{R}^n \times \mathbb{R} \mid \|x\|_2 \leq t\}$:

$$F(x, t) = -\log(t^2 - \|x\|_2^2), \qquad \text{with } \vartheta = 2. \tag{4.6}$$

The first case (4.4) leads to the logarithmic barrier function for linear programming, while the third (4.6) is important in another class of optimization problems known as *second-order cone programming* (the set K in this case is the Lorentz cone, known less formally as the "ice-cream cone"). The second case (4.5) is the one needed to construct primal barrier algorithms for semidefinite programming.

We use (4.5) to define the following barrier problem for (4.1):

$$\min_{X \succ 0} C \bullet X + \mu F(X) \;\text{ s.t. }\; A_i \bullet X = b_i, \; i = 1, 2, \ldots, m. \tag{4.7}$$

The dual barrier problem is derived similarly from (4.2):

$$\max_{S \succ 0, y} b^T y - \mu F(S) \;\text{ s.t. }\; \sum_{i=1}^{m} A_i y_i + S = C. \tag{4.8}$$

These barrier problems are useful not only in defining algorithms (see Section 5) but also in studying properties of the *central path*, which is the set of triples $(X(\mu), y(\mu), S(\mu))$ such that $X(\mu)$ solves (4.7) and $(y(\mu), S(\mu))$ solves (4.2). It can be shown that such solutions exist and that the path is smooth. For a proof of the following result, see Todd [**Tod01**, Section 5].

THEOREM 4.4. *Suppose that the feasible set for (4.1) is nonempty and contains an X with $X \succ 0$, and that the feasible set for (4.2) is nonempty and contains an element (y, S) with $S \succ 0$. Then there is a unique solution $X(\mu)$ of (4.1) and a unique solution $(y(\mu), S(\mu))$ to (4.2). Moreover, these solutions satisfy*

$$X(\mu)S(\mu) = S(\mu)X(\mu) = \mu I. \tag{4.9}$$

Finally, the central path, which contains $(X(\mu), y(\mu), S(\mu))$ for all positive μ, is differentiable with respect to μ.

Note that the points $(X(\mu), y(\mu), S(\mu))$ on the central path satisfy the following relations, for some $\mu > 0$.

$$A_i \bullet X = b_i, \quad i = 1, 2, \ldots, m, \tag{4.10a}$$

$$\sum_{i=1}^{m} A_i y_i + S = C \tag{4.10b}$$

$$XS = \mu I. \tag{4.10c}$$

5. Semidefinite Programming: Algorithms

Most of the recent research on algorithms for semidefinite programming has been based on the barrier problems (4.7) and (4.8) and the central path (4.10). We describe these interior-point methods in this section. Other important algorithmic work on special cases of (4.1) involves bundle methods for nonsmooth optimization. We do not describe these approaches here, but refer the interested reader to Helmberg and Rendl [**HR00**].

5.1. Primal and Dual Barrier Algorithms. Nesterov and Nemirovski [**NN94**] described algorithms that use Newton-like methods to solve approximately the primal barrier problem (4.7) for a fixed value of μ, then decrease μ and repeat the process. Since $F'(X) = -X^{-1}$, we have from first-order optimality that the solution of (4.7) satisfies

$$A_i \bullet X = b_i,\ i = 1, 2, \ldots, m, \qquad \sum_{i=1}^{m} A_i y_i + \mu X^{-1} = C.$$

Defining $S = \mu X^{-1}$, we can write these conditions as follows:

$$A_i \bullet X = b_i, \quad i = 1, 2, \ldots, m, \tag{5.1a}$$

$$\sum_{i=1}^{m} A_i y_i + S = C \tag{5.1b}$$

$$-\mu X^{-1} + S = 0. \tag{5.1c}$$

(Note that these conditions represent a nonlinear transformation of the central path conditions (4.10).) By applying Newton's method to the system (5.1), we obtain SQP steps for (4.7).

Nesterov and Nemirovski [**NN94**] show that this approach allows both "short-step" and "long-step" variants. In short-step methods, a single SQP step is taken for each value of μ, and μ is reduced by only a modest factor between iterations. In long-step algorithms, more ambitious reductions in μ are made, but several SQP steps are required. These steps may require "damping," that is, their length could be reduced to ensure that the next iterate X remains positive definite. In either case, the self-concordant nature of the barrier function F can be used to quantify the efficacy of the SQP steps, making it possible to derive upper bounds on the total number of steps required to reduce μ below a given threshold. (For the short-step approach, for a suitable starting point and a given threshold value ϵ, it requires $O(\sqrt{n}\log(1/\epsilon))$ iterations to achieve $\mu \le \epsilon$.) It follows that the total computational effort required is polynomial in the dimension n.

Similar approaches can be applied to the dual barrier problem (4.2). However, there may be a significant practical advantage to working with the dual formulation. When the data matrices A_i and C are sparse (as they often are in applications), the optimal S is usually sparse as well, because of the relationship

$$S = C - \sum_{i=1}^{m} A_i y_i. \tag{5.2}$$

On the other hand, the optimal X is dense in general. Several authors have exploited this property by designing dual barrier approaches for subclasses of the semidefinite programming; in particular, classes for which the constraints include specification of the diagonal of X or its trace. (Such constraints make it easy to identify a strictly feasible point for (4.2).) Burer, Monteiro, and Zhang [**BMZ02**] work with a parametrization of S that maintains positive definiteness, and use steepest descent to solve the resulting dual barrier subproblem. Fukuda, Kojima, and Shida [**FKS02**] apply a Newton or quasi-Newton method to the subproblem, and use an extrapolation based on preconditioned conjugate gradient to obtain a starting point for each new value of μ.

5.2. Primal-Dual Interior-Point Algorithms. Primal-dual interior-point algorithms generate a sequence of iterates (X, y, S) with positive definite X and S, such that X converges to a solution of (4.1) and (y, S) converges to a solution of (4.2). Path-following methods attempt to follow the central path to the solution, by finding approximate solutions of (4.10) for smaller and smaller values of μ. In the case of linear programming, primal-dual methods calculate Newton steps (3.2) for a system like (4.10) and take damped steps along these directions, requiring all iterates to remain inside some loose neighborhood of the central path. However, this approach cannot be applied directly to (4.10), because the domain and range spaces of the nonlinear system (4.10) are different. Since XS is nonsymmetric in general, the range space is $\mathbf{R}^m \times S\mathbf{R}^{n\times n} \times \mathbf{R}^{n\times n}$, while the domain is $S\mathbf{R}^{n\times n} \times \mathbf{R}^m \times S\mathbf{R}^{n\times n}$. We need to find an alternative definition of the central path in which the domain and range are equal before applying Newton's method. Note that (5.1) is one such redefinition. However it is not necessarily the best one because the smooth, mildly nonlinear equation $XS = \mu I$ has been replaced by an equation involving X^{-1}, which becomes highly nonlinear as X approaches singularity.

One of the first ideas in this vein, due to Alizadeh, Haeberly, and Overton [**AHO98**] was to replace (4.10c) by

$$\frac{1}{2}(XS + SX) = \mu I. \tag{5.3}$$

By applying Newton's method to (4.10a), (4.10b), and (5.3), and assuming that the current iterate (X, y, S) is feasible with respect to the conditions (4.10a) and (4.10b), we obtain the following linear system to be satisfied by the step $(\Delta X, \Delta y, \Delta S)$:

$$A_i \bullet \Delta X = 0, \quad i = 1, 2, \ldots, m, \tag{5.4a}$$

$$\sum_{i=1}^{m} A_i \Delta y_i + \Delta S = 0 \tag{5.4b}$$

$$\frac{1}{2}(X\Delta S + \Delta X S + S\Delta X + \Delta S X) = \mu I - \frac{1}{2}(XS + SX). \tag{5.4c}$$

A Lyapunov system involving eigendecompositions of X and S must be solved to find $(\Delta X, \Delta y, \Delta S)$, and the step may not be defined in general. Nevertheless, this approach has proved to be quite successful in practice.

A generalization of the symmetrization (5.3) is as follows:

$$\frac{1}{2}(PXSP^{-1} + P^{-T}SXP^T) = \mu I, \tag{5.5}$$

where P is a nonsingular $n \times n$ matrix (possibly depending on X and S). The class of methods resulting from this reformulation is known as the Monteiro-Zhang (MZ) family (see [**Mon97, Zha98**]). A number of algorithms were found after their initial discovery to belong to this family. One such is the Nesterov-Todd

(NT) direction [**NT97, NT98**], which is obtained by setting $P = W^{-1/2}$, where $W = X^{1/2}(X^{1/2}SX^{1/2})^{-1/2}X^{1/2}$. (This W is the unique positive definite matrix for which $WSW = X$.) The method of Nesterov and Todd arose from their work on self-scaled cones, which was designed as a vehicle for generalizing primal-dual interior-point methods for linear programming to other convex optimization problems involving a particular type of feasible cone. Renegar's monograph [**Ren01**] presents a beautiful explication of this theory.

The discussion above by no means exhausts the list of primal-dual methods that have been proposed and implemented. Todd's survey paper [**Tod99**] describes 20 directions.

Short-step and long-step variants of the primal-dual approaches mentioned above have been developed, and complexity results similar to those for the primal barrier approach have been obtained. There are also "potential-reduction" methods, which use search directions like those described in this section, but use a logarithmic potential function to choose the step length to be taken along these directions and to monitor progress.

Very good software for solving general semidefinite programming problems using primal-dual methods is available; the codes SeDuMi, SDPT3, and SDPA are widely used and can be located easily on the Web.

6. Conclusion

We have given a brief overview of two of the most active areas in continuous optimization, summarizing the main theoretical results and outlining some of the major algorithmic approaches to solving these problems. Our discussion has of necessity been incomplete. For additional information on nonlinear programming, we refer to the books of Bertsekas [**Ber99**] and Nocedal and Wright [**NW99**]. Additional information on semidefinite programming can be found in the sources mentioned in Section 4. Further developments can be expected in both areas as researchers rise to the challenges posed by larger and more complex applications.

Acknowledgments

I thank the organizers of the Short Course and the two referees of this note, who provided very helpful comments.

References

[AHO98] F. Alizadeh, J.-P. A. Haeberly, and M. L. Overton, *Primal-dual interior-point methods for semidefinite programming: Convergence rates, stability, and numerical results*, SIAM Journal on Optimization **8** (1998), 746–768.

[Ali91] F. Alizadeh, *Combinatorial optimization with interior point methods and semidefinite matrices*, Ph.D. thesis, University of Minnesota, 1991.

[Ali92] ———, *Optimization over positive semi-definite cone; interior-point methods and combinatorial applications*, Advances in Optimization and Parallel Computing (P.M. Pardalos, ed.), North–Holland, 1992, pp. 1–25.

[Ber82] D. P. Bertsekas, *Constrained optimization and lagrange multiplier methods*, Academic Press, New York, 1982.

[Ber99] D. P. Bertsekas, *Nonlinear programming*, 2nd. ed., Athena Scientific, Belmont, Mass., 1999.

[BHN99] R. H. Byrd, M. E. Hribar, and J. Nocedal, *An interior point algorithm for large scale nonlinear programming*, SIAM Journal on Optimization **9** (1999), no. 4, 877–900.

[BMZ02] S. Burer, R. D. C. Monteiro, and Y. Zhang, *Solving a class of semidefinite programs via nonlinear programming*, Mathematical Programming, Series A **93** (2002), 97–122.

[BN01] A. Ben-Tal and A. Nemirovski, *Lectures on modern convex optimization: Analysis, algorithms, and engineering applications*, MPS-SIAM Series on Optimization, SIAM, 2001.

[BNO03] D. P. Bertsekas, A. Nedić, and A. E. Ozdaglar, *Convex analysis and optimization*, Optimization and Computation Series, Athena Scientific, 2003.

[CGT91] A. R. Conn, N. I. M. Gould, and Ph. L. Toint, *A globally-convergent augmented Lagrangian algorithm for optimization with general constraints and simple bounds*, SIAM Journal on Numerical Analysis **28** (1991), 545–572.

[CGT92] A. R. Conn, N. I. M. Gould, and Ph. L. Toint, *LANCELOT: a Fortran package for large-scale nonlinear optimization*, Springer Series in Computational Mathematics, vol. 17, Springer-Verlag, 1992.

[Cla83] F. H. Clarke, *Optimization and nonsmooth analysis*, John Wiley, New York, 1983.

[Dan63] G. B. Dantzig, *Linear programming and extensions*, Princeton University Press, Princeton, New Jersey, 1963.

[Fay97a] L. Faybusovich, *Euclidean Jordan algebras and interior-point algorithms*, Positivity **1** (1997), 331–357.

[Fay97b] ———, *Linear systems in Jordan algebras and primal-dual interior-point algorithms*, Journal of Computational and Applied Mathematics **86** (1997), 149–175.

[FKS02] M. Fukuda, M. Kojima, and M. Shida, *Lagrangian dual interior-point methods for semidefinite programs*, SIAM Journal on Optimization **12** (2002), no. 4, 1007–1031.

[FM68] A. V. Fiacco and G. P. McCormick, *Nonlinear programming: Sequential unconstrained minimization techniques*, Wiley, New York, 1968, Reprinted by SIAM Publications, 1990.

[GMS02] P. E. Gill, W. Murray, and M. Saunders, *SNOPT: An SQP algorithm for large-scale constrained optimization*, SIAM Journal on Optimization **12** (2002), no. 4, 979–1006.

[Gri00] A. Griewank, *Evaluating derivatives: Principles and techniques of algorithmic differentiation*, Frontiers in Applied Mathematics, SIAM, Philadelphia, PA, 2000.

[Han77] S. P. Han, *A globally convergent algorithm for nonlinear programming*, Journal of Optimization Theory and Applications **22** (1977), 297–309.

[Hes69] M. R. Hestenes, *Multiplier and gradient methods*, Journal of Optimization Theory and Applications **4** (1969), 303–320.

[HR00] C. Helmberg and F. Rendl, *A spectral bundle method for semidefinite programming*, SIAM Journal on Optimization **10** (2000), 673–696.

[Joh48] F. John, *Extremum problems with inequalities as subsidiary conditions*, Studies and Essays: Courant Anniversary Volume (K. O. Friedrichs, O. E. Neugebauer, and J. J. Stoker, eds.), Wiley-Interscience, New York, 1948, pp. 187–204.

[Kar39] W. Karush, *Minima of functions of several variables with inequalities as side conditions*, Master's thesis, University of Chicago, Department of Mathematics, 1939.

[KT51] H. W. Kuhn and A. W. Tucker, *Nonlinear programming*, Proceedings of the Second Berkeley Symposium on Mathematical Statistics and Probability (Berkeley) (J. Neyman, ed.), University of California Press, 1951, pp. 481–492.

[Mar78] N. Maratos, *Exact penalty function algorithms for finite dimensional and control optimization problems*, Ph.D. thesis, University of London, 1978.

[Mon97] R. D. C. Monteiro, *Primal-dual path-following algorithms for semidefinite programming*, SIAM Journal on Optimization **7** (1997), 663–678.

[NN94] Yu. E. Nesterov and A. S. Nemirovskii, *Interior point polynomial methods in convex programming*, SIAM Publications, Philadelphia, 1994.

[NT97] Yu. E. Nesterov and M. J. Todd, *Self-scaled barriers and interior-point methods for convex programming*, Mathematics of Operations Research **22** (1997), 1–42.

[NT98] ———, *Primal-dual interior-point methods for self-scaled cones*, SIAM Journal on Optimization **8** (1998), 324–362.

[NW99] J. Nocedal and S. J. Wright, *Numerical Optimization*, Springer, New York, 1999.

[Omo89] E. O. Omojokun, *Trust-region algorithms for optimization with nonlinear equality and inequality constraints*, Ph.D. thesis, University of Colorado, Boulder, 1989.

[Ous00] F. Oustry, *A second-order bundle method to minimize the maximum eigenvalue function*, Mathematical Programming **89** (2000), 1–33.

[OW93] M. L. Overton and R. S. Womersley, *Optimality conditions and duality theory for minimizing sums of the largest eigenvalues of symmetric matrices*, Mathematical Programming **62** (1993), 321–357.

[Pow69] M. J. D. Powell, *A method for nonlinear constraints in minimization problems*, Optimization (R. Fletcher, ed.), Academic Press, New York, 1969, pp. 283–298.

[Pow77] M. J. D. Powell, *A fast algorithm for nonlinearly constrained optimization calculations*, Dundee Conference on Numerical Analysis (Dundee) (G. A. Watson, ed.), vol. 7, Springer-Verlag, June 1977.

[Ren01] J. Renegar, *A mathematical view of interior-point methods in convex optimization*, MPS-SIAM Series in Optimization, SIAM, 2001.

[Roc70] R. T. Rockafellar, *Convex analysis*, Princeton University Press, Princeton, N.J., 1970.

[Roc71] R. T. Rockafellar, *New applications of duality in convex programming*, Proceedings on the Fourth Conference on Probability, Brasov, Romania, 1971, pp. 73–81.

[Roc73] ———, *A dual approach to solving nonlinear programming problems by unconstrained optimization*, Mathematical Programming **5** (1973), 354–373.

[Roc76] R. T. Rockafellar, *Augmented Lagrangians and applications of the proximal point algorithm in convex programming*, Mathematics of Operations Research **1** (1976), 97–116.

[RW98] R. T. Rockafellar and R. J.-B. Wets, *Variational analysis*, Springer, Berlin, 1998.

[SBV02] D. Shanno, H. Y. Benson, and R. J. Vanderbei, *Interior-point methods for nonconvex nonlinear programming*, Computational Optimization and Applications **23** (2002), 257–272.

[Tod99] M. J. Todd, *A study of search directions in primal-dual interior-point methods for semidefinite programming*, Optimization Methods and Software **11** & **12** (1999), 1–46.

[Tod01] M. J. Todd, *Semidefinite optimization*, Acta Numerica **10** (2001), 515–560.

[VB96] L. Vandenberghe and S. Boyd, *Semidefinite programming*, SIAM Review **38** (1996), 49–95.

[Wae02] A. Waechter, *An interior-point algorithm for large-scale nonlinear optimization with applications in process engineering*, Ph.D. thesis, Department of Chemical Engineering, Carnegie-Mellon University, 2002.

[WB00] A. Waechter and L. T. Biegler, *Failure of global convergence for a class of interior point methods for nonlinear programming*, Mathematical Programming, Series A **88** (2000), no. 3, 565–587.

[Wil63] R. B. Wilson, *A simplicial algorithm for concave programming*, Ph.D. thesis, Graduate School of Business Administration, Harvard University, 1963.

[WMNO03] R. A. Waltz, J.-L. Morales, J. Nocedal, and D. Orban, *KNITRO-Direct: A hybrid interior algorithm for nonlinear optimization*, Technical report, Department of Electrical and Computer Engineering, Northwestern University, August 2003.

[WR96] S. J. Wright and D. Ralph, *A superlinear infeasible-interior-point algorithm for monotone nonlinear complementarity problems*, Mathematics of Operations Research **21** (1996), 815–838.

[Wri92] M. H. Wright, *Interior methods for constrained optimization*, Acta Numerica 1992, Cambridge University Press, Cambridge, 1992, pp. 341–407.

[WSV00] H. Wolkowicz, R. Saigal, and L. Vandenberghe (eds.), *Handbook of semidefinite programming*, Kluwer, 2000.

[ZBF+03] Z. Zhao, B. Braams, M. Fukuda, M. L. Overton, and J. Percus, *The reduced density matrix method for electronic structure calculations and the role of three-index representability conditions*, Technical report, Courant Institute of Mathematical Sciences, October 2003.

[Zha98] Y. Zhang, *On extending primal-dual interior-point algorithms from linear programming to semidefinite programming*, SIAM Journal on Optimization **8** (1998), 365–386.

Computer Sciences Department, University of Wisconsin, 1210 W. Dayton Street, Madison, WI 53706

E-mail address: swright@cs.wisc.edu

Index

Titles in This Series

61 **Serkan Hoşten, Jon Lee, and Rekha R. Thomas, Editors,** Trends in optimization (Phoenix, Arizona, January 2004)

60 **Susan G. Williams, Editor,** Symbolic dynamics and its applications (San Diego, California, January 2002)

59 **James Sneyd, Editor,** An introduction to mathematical modeling in physiology, cell biology, and immunology (New Orleans, Louisiana, January 2001)

58 **Samuel J. Lomonaco, Jr., Editor,** Quantum computation: A grand mathematical challenge for the twenty-first century and the millennium (Washington, DC, January 2000)

57 **David C. Heath and Glen Swindle, Editors,** Introduction to mathematical finance (San Diego, California, January 1997)

56 **Jane Cronin and Robert E. O'Malley, Jr., Editors,** Analyzing multiscale phenomena using singular perturbation methods (Baltimore, Maryland, January 1998)

55 **Frederick Hoffman, Editor,** Mathematical aspects of artificial intelligence (Orlando, Florida, January 1996)

54 **Renato Spigler and Stephanos Venakides, Editors,** Recent advances in partial differential equations (Venice, Italy, June 1996)

53 **David A. Cox and Bernd Sturmfels, Editors,** Applications of computational algebraic geometry (San Diego, California, January 1997)

52 **V. Mandrekar and P. R. Masani, Editors,** Proceedings of the Norbert Wiener Centenary Congress, 1994 (East Lansing, Michigan, 1994)

51 **Louis H. Kauffman, Editor,** The interface of knots and physics (San Francisco, California, January 1995)

50 **Robert Calderbank, Editor,** Different aspects of coding theory (San Francisco, California, January 1995)

49 **Robert L. Devaney, Editor,** Complex dynamical systems: The mathematics behind the Mandlebrot and Julia sets (Cincinnati, Ohio, January 1994)

48 **Walter Gautschi, Editor,** Mathematics of Computation 1943–1993: A half century of computational mathematics (Vancouver, British Columbia, August 1993)

47 **Ingrid Daubechies, Editor,** Different perspectives on wavelets (San Antonio, Texas, January 1993)

46 **Stefan A. Burr, Editor,** The unreasonable effectiveness of number theory (Orono, Maine, August 1991)

45 **De Witt L. Sumners, Editor,** New scientific applications of geometry and topology (Baltimore, Maryland, January 1992)

44 **Béla Bollobás, Editor,** Probabilistic combinatorics and its applications (San Francisco, California, January 1991)

43 **Richard K. Guy, Editor,** Combinatorial games (Columbus, Ohio, August 1990)

42 **C. Pomerance, Editor,** Cryptology and computational number theory (Boulder, Colorado, August 1989)

41 **R. W. Brockett, Editor,** Robotics (Louisville, Kentucky, January 1990)

40 **Charles R. Johnson, Editor,** Matrix theory and applications (Phoenix, Arizona, January 1989)

39 **Robert L. Devaney and Linda Keen, Editors,** Chaos and fractals: The mathematics behind the computer graphics (Providence, Rhode Island, August 1988)

38 **Juris Hartmanis, Editor,** Computational complexity theory (Atlanta, Georgia, January 1988)

ISBN 0-8218-3584-X
9 780821 835845
PSAPM/61